QUANTUM PHYSICS

FOR BEGINNERS

The Non-Scientist's Guide to the Big Ideas of Quantum Mechanics, with Key Principles, Major Theories, and Experiments Simplified

Second Edition

PANTHEON SPACE ACADEMY

QUANTUM PHYSICS
FOR BEGINNERS
The Non-Scientist's Guide to the Big Ideas of Quantum Mechanics, with Key Principles, Major Theories, and Experiments Simplified
Second Edition

Pantheon Space Academy

Printed Worldwide
Second Printing 2024
Second Edition 2024

QUANTUM PHYSICS

FOR BEGINNERS

INCLUDED

For additional resources and free downloads included with your book purchase, visit **www.pantheonspace.com/Qplus**

These include classic works like:

- ***The Theory of Relativity* by Albert Einstein**
- **The Origin of Quantum Theory by Max Planck**
- **Eight Lectures delivered at Columbia by Max Planck**
- **The Nature of the Physical World by Sir Arthur Stanley Eddington**

You can also download my extended glossary, which offers in-depth explanations and hundreds of terms to deepen your understanding of quantum physics. All of these digital files, along with even more exclusive content, come with this second edition of the book. I encourage you to explore these materials and enhance your learning experience right away!

Be notified of new releases and special promotions by following my author page at **www.amazon.com/author/xyz**

Keep the conversation going by joining **www.facebook.com/pantheonspace**

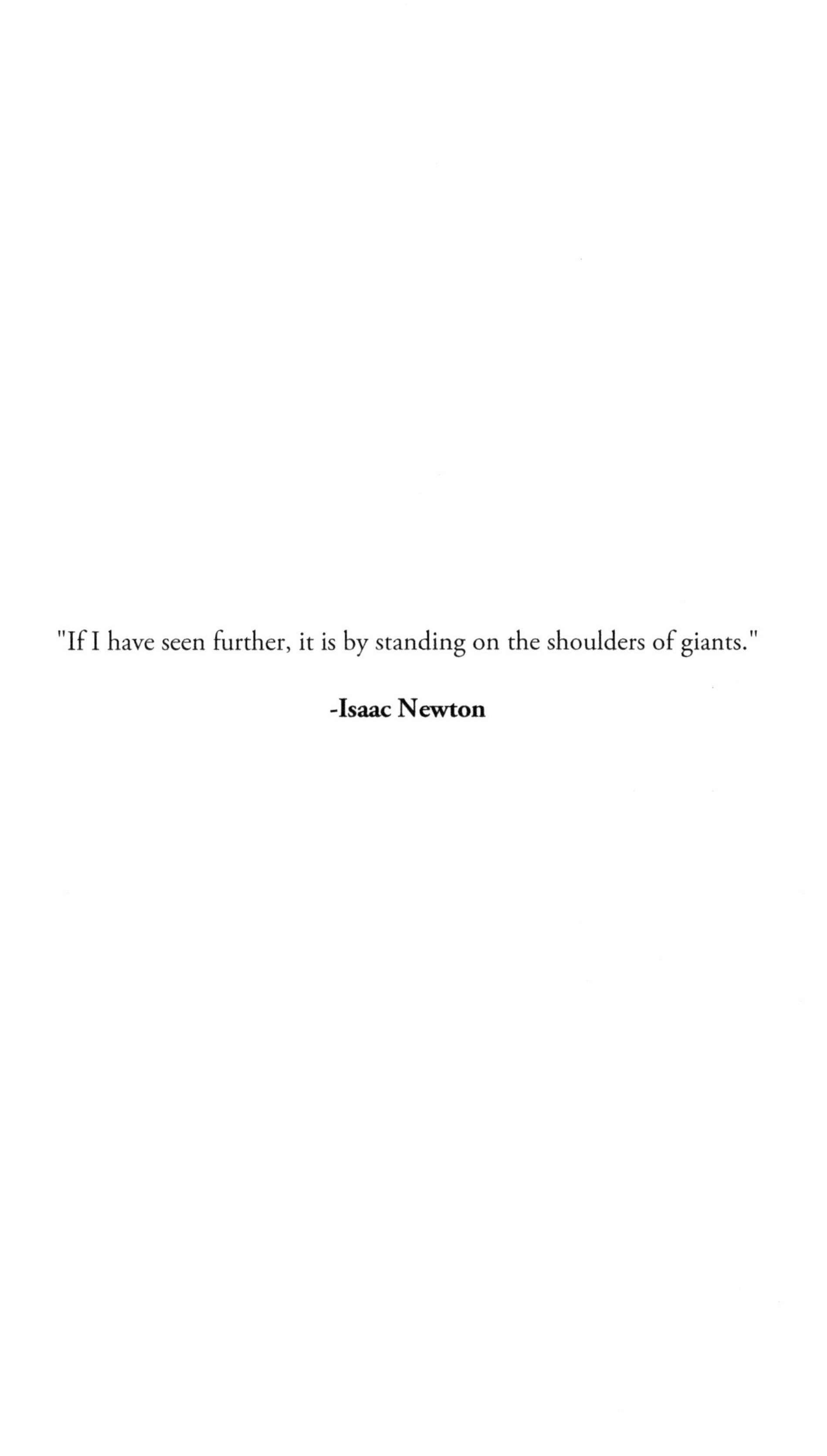

"If I have seen further, it is by standing on the shoulders of giants."

-Isaac Newton

TABLE OF CONTENTS

Preface

When the first edition of this book was released, I was thrilled by the response. Readers from all walks of life shared their thoughts, questions, theories, and inventions. Among the most frequent requests was for more intimate details about the people behind the discoveries—those giants of science whose perseverance and creativity shaped the quantum landscape. Many wanted to know not just about their theories but about the personal stories that drove their groundbreaking work.

It became clear to me that something was missing from the original book: the human side of these great physicists—their struggles, triumphs, and how their personal journeys influenced their scientific discoveries. I've always been fascinated by how adversity shapes a person's philosophy and methods. For these physicists, the struggle wasn't just a backdrop—it was a driving force behind their ability to push boundaries and redefine the laws of nature. Their ability to overcome can inspire us all.

With this second edition, I aim to bridge that gap by introducing *The Giants of Science*, a *"mini-book"* dedicated to the personal lives and experiences of ten of the most influential minds in quantum physics. Through their stories, you'll see that science is not just a body of knowledge but a deeply human pursuit, filled with moments of failure, determination, and brilliance. By understanding the personal contexts in which these theories were born, you'll gain a deeper and more memorable understanding of quantum mechanics.

In addition to this new content, I've also included special access to free ebooks featuring works by Albert Einstein, Max Planck, and others, to further enrich your understanding of these pioneers. You'll find an **Included** section with a link to these materials at the beginning and conclusion of this book, allowing you to dive even deeper into their thoughts, theories, and contributions.

As you read through this edition, I hope you're as inspired by these stories as I am. There's a pattern to be found in the lives of these giants—Can you spot any standout moments or common threads that tie their journeys together? Their stories may hold lessons that extend far beyond the laboratory, offering insights into science and the human spirit.

Would you rather dive directly into the theories and experiments covered in the first edition? I invite you to start with Part 2 on page 85. When you're ready to explore the human stories behind the discoveries, you can return to Part 1 and connect with the personal stories that shaped quantum physics. Whichever path you choose, I hope this edition brings you new insights and a deeper connection to the brilliant minds that revolutionized our understanding of the universe.

PART 1

THE GIANTS OF SCIENCE

How personal challenges and triumphs shaped the scientific breakthroughs of ten iconic quantum physicists

INTRODUCTION

SECOND EDITION, PART 1

"*Somewhere, something incredible is waiting to be known.*"

—Carl Sagan

When we think of scientific breakthroughs, it's easy to picture them as isolated achievements—equations on a chalkboard, theories in a vacuum. But science, like life, is far more complex. Just as threads in a fabric are inseparable, so are the personal and professional lives of the great physicists who shaped quantum mechanics. These giants of science weren't just brilliant minds; they were sons, fathers, friends, mentors—people navigating the same human challenges we all face.

Consider the world they lived in. These scientists worked during times of immense political upheaval, moral dilemmas, and societal transformation. World wars, exiles, and ideological battles influenced their paths as much as their scientific education. Many grappled with more than just the mysteries of the quantum world—they wrestled with the ethical importance of their discoveries. How would their work impact society? How would they balance scientific exploration with the potential consequences of their breakthroughs? These were not just academic questions, but deeply personal ones.

Their stories extend far beyond the equations and experiments. Relationships—whether filled with collaboration or conflict—played a crucial role in their discoveries. Some forged friendships that fueled new ideas, while others clashed, sharpening their own theories in the process. Together, their combined efforts pushed the boundaries of human knowledge. It reminds us that science isn't just about individual genius but the collective struggle and success of many minds working toward a common goal.

Behind every major discovery were moments of uncertainty, doubt, and even failure. Yet, what united these physicists was their shared determination to keep moving forward. They faced the same emotions we all do: fear of failure, frustration, hope, and ambition. But in the end, their persistence, curiosity, and ability to learn from setbacks paved the way for some of the greatest advancements of our time.

So, as you explore the lives of these pioneers, consider the world they lived in, the relationships they formed, and the challenges they overcame. Their stories hold lessons that resonate far beyond the laboratory, offering insights into the nature of the universe and the nature of the human spirit. What drove them to question the fabric of reality itself? It's time to turn the page and explore the minds that have changed our understanding of reality.

Max Planck

Tradition Meets Revolution

A conservative upbringing and initial obedience to classical physics set the stage for an unexpected encounter with groundbreaking concepts. Born in 19th-century Germany to a family that valued stability and tradition, these standards deeply influenced Planck's intellectual path. He embraced classical physics with the belief that nature followed predictable, continuous laws. His respect for the past caused him to seek solutions within the established rules. But, a nudge from black-body radiation would challenge him out of his comfort zone.

Planck's Journey into Quantum Mechanics

Berlin was a city of rapid industrialization and a hub of scientific innovation. Max immersed himself in this academic atmosphere. Yet, as much as he was shaped by tradition, the mysterious black-body radiation problem forced him into new territory. The classical theories that had long guided physicists couldn't explain the observed spectrum of radiation emitted by heated objects. Planck was at a crossroads: remain anchored in tradition or venture into the unknown.

In his groundbreaking work on black-body radiation, Planck proposed that energy is quantized—that it comes in discrete packets, or "quanta," rather than being continuous. At first, even Planck saw this as a mere mathematical fix rather than a fundamental shift in physics. He didn't know it yet, but his equation, $E=h\nu$, where h is now known as Planck's constant, would lay the foundation for quantum theory.

The Introduction of A Radical Idea

Planck's constant, h, emerged from his efforts to solve the black-body radiation puzzle. At the time, classical physics predicted that energy emitted at short wavelengths would approach infinity—a result known as the "ultraviolet catastrophe." This paradox, where theory predicted infinite energy emission at high frequencies (such as ultraviolet light) from heated objects, clashed with observed results and underscored the limitations of classical physics. Planck, desperate for a solution, proposed that the energy of atomic oscillators—vibrating particles within atoms responsible for emitting radiation—could only take on discrete values proportional to the frequency of radiation, which refers to the energy carried by electromagnetic waves like light. This radical idea was the birth of quantum theory.

At first, Planck didn't fully grasp the depth of his discovery. He was a reluctant revolutionary. Yet, his constant became a cornerstone of quantum mechanics, reshaping our understanding of energy and matter. His once-hesitant idea sparked the beginning of a scientific revolution.

Personal Challenges and Scientific Triumphs

While Planck was wrestling with the implications of his quantum breakthrough, his personal life was marked by profound tragedies. In 1909, Planck's beloved wife, Marie, died suddenly, leaving him devastated. Music, a lifelong passion, became a source of comfort during these dark times. Planck was an accomplished pianist, and his home was often filled with the sounds of Bach and Beethoven. In those moments, the lines between art and science blurred, and Planck found peace in the structure and beauty of music, much like the order he sought in the universe through physics.

Tragedy did strike again during World War I. Planck lost two of his sons—Karl, who died in battle, and Erwin, who was executed for his role in an assassination attempt on Hitler. Amid these personal losses, Planck's work became both an anchor and a distraction. His ability to continue pursuing science despite such immense grief revealed a depth of resilience that would inspire us more than 75 years later.

Collaboration

Max Planck was not alone in his struggles or triumphs. His relationship with other scientists became crucial to the advancement of quantum theory. Initially hesitant about the importance of his discovery, Planck's work caught the attention of a young Albert Einstein. Einstein was fascinated by Planck's quantum idea, built on his work to explain the photoelectric effect, demonstrating that light itself could be quantized into particles called photons. This discovery earned Einstein the Nobel Prize and solidified Planck's constant as a central element of quantum theory.

The friendship between Planck and Einstein was more than just intellectual. They shared a deep respect for one another's minds, often playing music together in Planck's Berlin home. Planck, on the piano, and Einstein, on the violin, would lose themselves in the melodies of Beethoven. These spontaneous duets symbolized the harmony between their academic and personal lives, a blend of creativity and science that left a lasting impact on both men.

His Revolutionary Role

Despite his groundbreaking contributions, Planck wrestled with his place in the revolution he had started. For a long time, he viewed his quantization of energy as nothing more than a mathematical tool to solve the black-body radiation problem. Initially skeptical of the radical shift toward quantum mechanics, Max sought ways to connect his findings with classical physics. Perhaps his deep respect for tradition kept him anchored to the familiar boundaries of classical theory, delaying his recognition of the profound impact his work would have on the future of physics.

The scientific community, too, was slow to accept Planck's ideas. Many traditional physicists resisted the idea of quantized energy, clinging to the belief that energy should remain a continuous, measurable entity. Planck and colleagues found it difficult to reconcile the unpredictability and probabilistic nature of quantum mechanics with the orderly, deterministic universe they already believed in. Despite the significant criticism and pushback, Planck continued to build upon his theories.

His reluctance was further tested by the rising tide of younger physicists, like Niels Bohr and Werner Heisenberg, who eagerly embraced quantum theory. Though initially hesitant, Planck became a mentor to these younger scientists, guiding them as they expanded on his ideas. His willingness to engage with and support them truly shaped the development of quantum theory. He provided the environment that allowed these young men to challenge classical physics and push science into uncharted territories.

To an outsider, the chalk-dusted equation conversations may have seemed mundane, but to those in the room, they were intense and filled with the excitement of discovery and debate. But Planck, always composed and reflective, would listen carefully; his humility and intellectual curiosity allowed him to absorb their ideas, learn from them, and ultimately adapt his own views. Bohr, in particular, credited Planck's work on quantization as the foundation for his own model of the atom. Heisenberg, inspired by Planck's constant, would develop the uncertainty principle, a cornerstone of quantum mechanics that expanded on the probabilistic nature of the quantum world. More on these men later.

Legacy of Planck's Constant

We now know that, over time, Planck came to accept the broader implications of his work. His constant became integral to the development of quantum mechanics, shaping everything from Heisenberg's uncertainty principle to Schrödinger's equation. More than just a scientific tool, Planck's constant symbolized a cultural shift in the scientific world—from the deterministic, predictable laws of classical physics to the probabilistic, uncertain realm of quantum mechanics.

His discovery challenged the Newtonian notion of a clockwork universe, where everything could be predicted if one had enough information. Instead, Planck introduced the concept of uncertainty, suggesting that nature operates on probabilities, not certainties. This idea revolutionized not just science but also philosophy, as it called into question humanity's ability to fully understand the universe. Planck's constant became a symbol of the limits of human knowledge, influencing physics and discussions in philosophy, art, and literature, where the concept of uncertainty resonated with a world grappling with modernity and change.

Planck's path from a classical physicist to the father of quantum theory is a testament to the power of open-mindedness and resilience. He faced immense personal loss, professional skepticism, and internal conflict yet emerged as a pioneer in modern physics, highlighting the excitement that great minds must face when their ideas outgrow their original intentions. His legacy, both in science and in life, reminds us that the greatest breakthroughs often come from those willing to question the world around them—no matter the cost.

Final Thought

Max Planck's life was marked by an extraordinary balancing act between tradition and innovation, personal grief, and professional triumph. His perseverance and humility laid the foundation for quantum mechanics, influencing generations of scientists. He taught us that while personal struggles may seem overwhelming, they can also be the source of profound growth and discovery. Behind the equations and constants lies a powerful human story—one of love, loss, resilience, and, ultimately, revolution.

By embracing change and nurturing new ideas, Planck set the stage for future generations of scientists to expand upon his work and further advance our understanding of the natural world. Next, we explore the life and contributions of Albert Einstein, who went from a struggling student to a science icon.

ALBERT EINSTEIN

THE RELUCTANT GENIUS

Albert Einstein's transformation from an average student to a renowned figure in the field of science is marked by unexpected developments. Often remembered as one of the greatest minds in history, Einstein's early academic struggles nearly derailed his educational prospects. His path was anything but smooth, yet these challenges played a crucial role in shaping the icon of modern physics. By exploring his early years, we develop a deeper understanding of how challenges and non-traditional decisions paved the way for significant achievements. In this section, I spotlight the importance of resilience, creativity, and the ability to think outside the box in pursuit of your goals.

Patent Clerk to Science Icon

When people think of Albert Einstein, they often picture the wild-haired, playful genius with a twinkle in his eye, delivering profound philosophical insights and quirky humor. Yet, few know that Einstein's life was fraught with personal struggles, academic failures, and self-doubt. His early student life was far from stellar. In his teenage years, he faced significant difficulties in school, where rigid teaching methods left him disconnected and unmotivated.

Despite his fascination with mathematics and physics, many of his teachers saw him as a mediocre student. He clashed with educators who valued conformity over curiosity, leading to poor grades and disciplinary issues. This frustration reached a breaking point when Einstein dropped out of school at the age of 15. For a young Albert, the structured environment of traditional education felt suffocating, and his rebellious nature often put him at odds with his teachers.

This period of struggle is familiar to young readers today—many of whom may feel misunderstood or restricted by traditional systems; I know I did. Yet, rather than seeing these obstacles as impossible, Einstein's story serves as a reminder that these setbacks can become opportunities for self-discovery. His experience in school shaped his determination to find his own path, one where creativity and intellectual freedom took priority over rigid expectations.

After dropping out of school, Einstein faced an uncertain future. He eventually completed his education in Switzerland, but his eccentric methods and clashes with professors—notably Heinrich Weber—left him without academic job prospects. Despite his clear brilliance, Einstein struggled to find his place within the scientific establishment. Unable to get a teaching or research position, he took a job at the Swiss Patent Office. For many, this would have been seen as a failure or an unwanted detour away from his scientific ambition, but for Einstein, it became an unlikely source of inspiration.

Now at work, Einstein reviewed and analyzed patent applications, often involving mechanical inventions and electrical devices. This seemingly dull work required meticulous attention to detail, but it also allowed him the freedom to ponder bigger questions about physics. Without the pressures of the old guard, Einstein was able to develop some of his most remarkable ideas in relative solitude. The patent office became a fertile ground for his intellectual pursuits, where his imagination could roam freely. It's a clear example of how unrelated jobs can contribute to one's growth, demonstrating that valuable insights can come if you have the right mindset.

One of the best lessons from Einstein's time at the patent office is the importance of unconventional thinking. Rather than adhering to the traditional academic path, Einstein used his unique situation to his advantage. His ability to think outside the box was directly influenced by his experience analyzing inventions, where he learned to look at problems from new angles. This period of his life reveals that no experience is ever wasted, and amazing ideas can emerge from any situation or personal technique.

Sailing Through Discovery

Einstein enjoyed sailing on lakes, particularly in Switzerland and Germany. One of his favorite places to sail was on Lake Geneva in Switzerland, where he spent much of his time during his years in Zurich and later visits. He also enjoyed sailing on Lake Constance, which borders Germany, Switzerland, and Austria. Sailing was a peaceful retreat for him, providing moments of reflection away from the pressures of academia. His time on the water was often spent in quiet contemplation, aligning with his love for solitude and his need to escape from the intense focus of his scientific work.

Some of his best thought experiments were during this refuge in the still waters. Away from the clutter of equations and the noise of lecture halls, For Einstein, sailing wasn't just an escape; it was a vital part of his creative process. He would often lose himself in deep contemplation of nature's secrets. As he sailed, he would envision light not just as a wave but also as discrete particles, and ponder how space and time were not fixed absolutes, but relative to the observer's motion. These moments gave him clarity before his most productive year in physics.

The "Annus Mirabilis"

Einstein's "Annus Mirabilis," or "Miracle Year," of 1905 was an exciting turning point in his career. In a single year, he published four groundbreaking papers that would forever change the landscape of physics. These papers covered topics ranging from Brownian motion to mass-energy equivalence, each of which contributed to the foundation of modern quantum theory. This period of Einstein's life highlights how practice and hyperfocus can lead to transformative shifts in your potential.

One of the most significant papers from this year introduced his special theory of relativity, which challenged long-standing notions of space and time. In a shocking leap, Einstein proposed that time and space were not absolute but relative to the observer, altering the established way scientists understood the universe. This theory would later be shown in his famous equation E=mc2, which proved the equivalence of energy and mass. This principle would have profound implications for everything from nuclear energy to cosmology.

At the same time, Einstein's work on the photoelectric effect revealed his ability to think outside the scientific norms of the time. In this paper, Einstein proposed that light behaves not only as a wave, as was traditionally thought, but also as particles or "quanta." This groundbreaking idea provided the first actual proof of the quantum nature of light, challenging the well-established wave theory of light. His work on the photoelectric effect was significant in developing quantum mechanics, earning him the Physics Nobel Prize in 1921.

During this period, Einstein's work began to attract attention from noteworthy figures in the quantum community, particularly Max Planck. Planck, the father of quantum theory, recognized the importance of Einstein's ideas and shared his papers with enthusiasm to help them gain wider acceptance. This collaboration between two of the most brilliant minds in physics was not just a professional alliance—it was a personal connection that provided Einstein with the validation and encouragement he desperately needed. For young scientists, the story of Planck's support serves as a reminder that even the greatest minds benefit from mentorship and collaboration.

The Development of Einstein's Theory

Einstein's discovery of the photoelectric effect was one of his most defining contributions to quantum physics, highlighting his innovative thinking and problem-solving abilities. At a time when the wave theory of light was widely accepted, Einstein proposed that light could also behave as particles, or photons. This radical idea forced scientists to reconsider the nature of light and marked a pivotal moment in the development of quantum mechanics.

To support his theory, Einstein relied on rigorous experimentation. One key experiment involved measuring the kinetic energy of electrons emitted from metal surfaces when exposed to light. This experiment showed that increasing the brightness of light increased the number of emitted electrons, but not their energy. On the contrary, increasing the frequency of light heightened the kinetic energy of the electrons without changing their quantity. These findings provided hard evidence for the existence of photons and confirmed Einstein's theory.

The significance of the photoelectric effect went beyond advancing quantum theory—it also demonstrated Einstein's ability to challenge established norms and think creatively about scientific problems. His work intersected with peers like Niels Bohr, whose atomic model drew upon Einstein's findings to explain energy levels in atoms. The collaboration between Einstein, Planck, and Bohr underscores the importance of intellectual dialogue in scientific discovery, showing that even the most extreme ideas are built on the contributions of many.

Final Thoughts

What are your hobbies? On quiet afternoons, Einstein would often retreat to his sailboat, gliding over the still waters, the rhythm of waves splashing against the hull. There, with the wind in his wild hair, his mind would wander freely, drifting between the physical world and the abstract universe of his thoughts. In these serene moments, with the sun sinking low on the horizon, some of his most profound ideas began to take shape. Much like the music he adored, particularly his treasured violin, these moments offered him solace, clarity, and a break from the structured complexities of his work. Einstein's genius wasn't limited to the chalkboard or lecture hall—it thrived in the open air, where life itself became his inspiration.

Even with his huge success, Einstein remained a humble figure. He would joke with friends, his eyes bright, insisting that his achievements were the result of curiosity and stubbornness rather than brilliance. Whether playing violin with Max Planck or walking home after a day's work, he seemed more in awe of the universe's mysteries than his role in unlocking them. To be with Einstein in these moments was to witness someone who saw the world not just through the lens of a scientist but with the heart of a dreamer. It was this childlike wonder that kept him grounded, always searching, always learning.

Niels Bohr

The Gentle Giant of Quantum Physics

Niels Bohr was more than just a pioneer of quantum mechanics—he was a physicist, philosopher, mentor, and deeply compassionate human being. Bohr's thoughts on ethics, his love for music, and his curiosity about the nature of reality shaped the field of physics and the lives of the people who crossed his path. In Bohr's world, scientific inquiry wasn't confined to the laboratory; It extended into the ethical and philosophical aspects that define what it means to be human.

Born in 1885 in Copenhagen to a close-knit Danish family known for intellectual curiosity and civic responsibility, Bohr was shaped by the ideals of his upbringing. His father, Christian Bohr, was a professor of physiology, and his mother, Ellen Adler Bohr, came from a wealthy and influential Jewish family. Bohr grew up in a home filled with discussions about politics, ethics, and the natural world, creating the foundation for his lifelong quest to understand both the universe and humanity's role within it.

Bohr was a man of great humility and kindness. His house in Copenhagen became a refuge for scientists from all over the world, hosting lively debates over quantum mechanics, followed by casual conversations over tea. Bohr, always the gracious host, made sure no one ever felt inferior, no matter how towering his reputation became. He once said, "Every great and deep difficulty bears in itself its own solution. It forces us to change our thinking in order to find it." This quote reflects how Bohr approached every problem in life and science—with patience, open-mindedness, and a deep sense of ethical purpose.

A Pioneer of Quantum Theory

Bohr's intellectual legacy is inseparable from his groundbreaking work on atomic structure and quantum theory. Early in his career, Bohr built on Ernest Rutherford's discovery of the nucleus to develop the Bohr model of the atom in 1913. In this model, he proposed that electrons travel in specific orbits around the nucleus and that their energy levels are quantized. This idea—electrons jumping between energy levels by absorbing or emitting photons—marked a significant leap in our understanding of the atom and laid the foundation for quantum mechanics.

Einstein's work on the photoelectric effect in 1905 helped establish that light can exhibit both wave-like and particle-like properties. However, Niels Bohr extended this concept beyond light, generalizing it to all particles, including electrons. It's known as the complementarity principle, which he developed in the late 1920s as part of the Copenhagen interpretation of quantum mechanics. This principle suggested that particles like electrons can exhibit both wave-like and particle-like behavior, but not at the same time—it depends on how the observer measures them. The idea was that the wave and particle nature are complementary aspects of the same reality, not contradictions.

So, to clarify:

- Einstein first introduced the idea that light (electromagnetic radiation) could behave as both particles (photons) and waves. This was an important breakthrough in quantum theory, for which Einstein received the Nobel Prize in 1921.
- Bohr then generalized this concept with his principle of complementarity, applying it not just to light, but to all particles (like electrons). Bohr's interpretation of quantum mechanics

emphasized that the behavior of particles depends on how they are observed—either as particles or as waves—but these are complementary and not mutually exclusive realities.

The principle of complementarity can be defined as the quality of being different but useful when combined, which is one of the cornerstones of quantum mechanics. It also became a philosophical framework for Bohr, and it reflects how he approached life. He believed that opposites can coexist in harmony—whether it's particle and wave behavior or peace and conflict in human society. Bohr's ability to hold multiple perspectives at once made him a groundbreaking scientist and a deep thinker about the world's big questions.

Mentor and Friend: Bohr's Legacy as a Teacher

One of the defining qualities of Bohr's career was his role as a mentor. He didn't just teach science—he taught how to be a scientist. Bohr's home was a gathering place for the brightest young minds in physics, and many of his students went on to become legends in their own right. People like Werner Heisenberg, Wolfgang Pauli, and Enrico Fermi were drawn to Bohr not just for his intellectual brilliance but for his personal warmth and humility.

Bohr had an open-door policy at his Copenhagen Institute for Theoretical Physics, where he would spend long hours discussing the most complex ideas with both seasoned scientists and students alike. His conversational style was gentle yet probing, always encouraging others to think deeper. Paul Dirac, a Nobel laureate and one of the key figures in quantum mechanics, once said, "To enter Bohr's institute was like going to the university of the future." Bohr didn't treat his students as inferiors; he treated them as equals, believing that great ideas could come from anyone.

His mentoring was deeply personal. Bohr developed lifelong friendships with many of his colleagues, but his relationship with Werner Heisenberg was particularly special. Heisenberg often referred to Bohr as a father figure, and their bond went beyond physics. Together, they tackled some of the most challenging questions in quantum theory. Still, their relationship wasn't always smooth—especially during World War II, when their opposing political positions strained their once-strong friendship. Yet, even as their disputes became more personal, Bohr's compassion and deep care for Heisenberg as a person never waned.

Even in the face of personal conflict, such as his strained relationship with Heisenberg during the war, Bohr never abandoned his belief in the power of dialogue. Though their differences tested their friendship, Bohr's dedication to open communication remained a guiding principle. 'I thought of him as my son,' Bohr reportedly said, illustrating the deep connection he felt with those he mentored, even during difficult times.

Bohr often said that the role of a teacher was not to provide answers but to help students ask better questions. He believed that the exchange of ideas, no matter how simple or complex, was the true foundation of scientific progress. 'Every great discovery begins with a conversation,' Bohr once remarked, a sentiment that defined his approach to mentoring and shaped the future of quantum mechanics. In Bohr's view, science was never a solo journey—it was a collective ambition, where collaboration and mutual respect were just as important as intellect.

War, Ethics, and Scientific Responsibility

The ethical dilemmas Bohr faced during World War II tested his deeply held beliefs about the role of science in society. As the Nazi regime spread across Europe, Bohr found himself at a crossroads. His Jewish heritage made him a target of persecution, and Denmark fell under Nazi occupation. Bohr was forced to flee to Sweden and, eventually, to the United States, where he became involved in the Manhattan Project—the secretive effort to develop the atomic bomb. His participation in the project was never without internal struggle. On the one hand, the atomic bomb represented a monumental leap in scientific understanding. On the other, it posed an existential threat to humanity.

Though Bohr contributed to the project by sharing his deep knowledge of atomic theory, he was never comfortable with the military use of nuclear energy. He believed that science should unite nations, not divide them. Even during the Manhattan Project, Bohr advocated for open dialogue between countries, including the Soviet Union, believing that the only way to avoid a catastrophic arms race was through transparency and cooperation. His talks with Robert Oppenheimer and other well-known scientists about the moral implications of their work planted seeds of doubt in many minds. He foresaw the dangers of nuclear weapons and spent much of his post-war life advocating for nuclear disarmament and the peaceful use of atomic energy. Figures like Leo Szilard, who had initially advocated for the project, became opponents of its use.

Bohr's ethical stance wasn't simply about avoiding destruction—it was about ensuring that science remained a force for good. He famously met with world leaders. Unlike many of his peers, Bohr was outspoken about these concerns. He urged world leaders Winston Churchill and Franklin D. Roosevelt to consider the long-term consequences, urging them to use the knowledge gained from the atomic bomb for peace rather than war. Despite the resistance he faced, Bohr never wavered in his commitment to international scientific cooperation, understanding that the long-term implications of nuclear energy extended far beyond national borders.

Life Beyond the Lab: Music, Skiing, and Family

What many people don't know about Niels Bohr is how rich his life was beyond the confines of the laboratory. Bohr was not only a physicist but also an avid skier and a music lover. He found in these pursuits a much-needed balance to the pressures of his work. He often took time off to ski in the mountains, clearing his mind and finding peace in nature. These moments of quiet reflection gave him the mental clarity he needed to solve some of the most challenging problems in physics.

Bohr's love for music also influenced his thinking. He would often compare a well-constructed scientific theory to a beautiful symphony—both required balance, harmony, and an appreciation for complexity. Much like Einstein, who found inspiration in the violin, Bohr found that music stimulated his creative thinking. He believed that science and the arts were interconnected, each offering different yet equally profound insights into the nature of reality.

But more than anything, Bohr was a family man. His wife, Margrethe, was his lifelong partner and friend. She played an instrumental role in his work, often reading his papers and discussing his ideas. Their home was filled with amazing discussions and warm family gatherings, where their six children grew up in an environment that valued curiosity and compassion. Bohr often credited his wife and family with giving him the emotional strength necessary to face all the challenges.

A Legacy of Humanity in Science

Let this be a reminder that the pursuit of knowledge is never isolated from the responsibility of how that knowledge is used. Niels's contributions to quantum mechanics changed the landscape of modern physics, but it was his deep moral compass and care for humanity that made him truly extraordinary. His belief in the power of collaboration, humility, and transparency continues to resonate with scientists today, many of whom look to him as a model for navigating ethical challenges.

In his later years, Bohr became a leading advocate for Atoms for Peace, a movement aimed at using nuclear technology for peaceful purposes. His speeches at international conferences emphasized that scientific progress must be paired with global cooperation, a message that remains relevant in today's world. Bohr's efforts didn't stop in the laboratory—he was a visionary for a future where science could be a tool for diplomacy and peace.

Bohr's life teaches us that being a scientist is not just about uncovering the secrets of the universe—it's about caring for the people around you and using knowledge to improve the world. His gentle spirit, combined with his intellectual fearlessness, made him a giant in both science and humanity. For young readers, Bohr's journey is an invitation to pursue their passions, ask difficult questions, and remember that no discovery is too small to impact the world.

WERNER HEISENBERG

THE PHILOSOPHER OF PHYSICS

Werner Heisenberg was a man of many dimensions—an accomplished musician, a philosopher, a mentor, and an innovator who changed the landscape of quantum mechanics forever. Known primarily for his famous uncertainty principle, Heisenberg's life was much more than what he contributed to science. His love for music, his role as a teacher, and his resilience in rebuilding post-war Germany shaped his legacy and the world of physics forever.

Born in 1901 in Würzburg, Germany, Werner Heisenberg grew up in an academic household. His father, August Heisenberg, was a professor of classical languages, and his mother, Annie Wecklein, came from a family with a long tradition of education. The intellectual atmosphere of Heisenberg's childhood, combined with the turmoil of two world wars, shaped his path. From a young age, Heisenberg was fascinated by science's precision and the philosophical questions that came with it.

A Source of Inspiration

Heisenberg's love for music was a constant throughout his life, one that often contributed to his scientific pursuits. He was an accomplished pianist, and like his contemporary Albert Einstein, Heisenberg found that music offered a kind of mental shift. It wasn't just a hobby for him but a way of thinking. Heisenberg once said, "Music and the rhythm of its compositions have always had an influence on my scientific ideas."

He would often sit at the piano and play pieces by Bach and Beethoven, using the structure and harmony of music as a metaphor for his work in quantum mechanics. Just as music required creativity and strict adherence to form, so did Heisenberg's approach to physics. He saw the principles of music as reflecting the same balance between order and unpredictability that he loved to explore in his work. Music gave him a space to reflect on the complexities of nature, offering him moments of clarity that fed back into his groundbreaking discoveries.

Heisenberg's colleagues often said his love for music spilled over into his work. During collaborative discussions with other physicists, he would draw parallels between a well-constructed scientific theory and a symphony. In his eyes, both required a perfect balance of creativity and structure. Wolfgang Pauli, one of Heisenberg's closest collaborators, recalled how Heisenberg would often punctuate intense scientific debates by sitting at the piano, allowing his mind to work through problems via the melody. This blend of art and logic became a hallmark of Heisenberg's approach to physics.

The Birth of the Uncertainty Principle

Heisenberg's most famous contribution to quantum mechanics is undoubtedly the uncertainty principle. First formulated in 1927, the principle states that certain pairs of physical properties, like position and momentum, cannot both be known to absolute precision at the same time. This idea turned the classical notion of a deterministic universe on its head and introduced the concept of fundamental limits to what we can know about the physical world.

But Heisenberg's path to this revolutionary idea wasn't very straightforward. In the early 1920s, the world of physics was on the edge of a major disruption. Classical physics, which had dominated for centuries, was being challenged by the strange, probabilistic nature of quantum mechanics. Heisenberg, still in his twenties, was at the forefront of this shift. His early work in matrix mechanics—a mathematical framework for quantum theory—had already established him as a rising star. It was his realization that nature itself was inherently uncertain that solidified his place in the history books.

Mathematically, this is expressed as:

$$\Delta x \cdot \Delta p \geq \frac{h}{4\pi}$$

In this equation:

- Δx represents the uncertainty in the particle's position, indicating how precisely we know where it is located.
- Δp is the uncertainty in the particle's momentum, describing how precisely we know its motion and speed.
- The product $\Delta x \cdot \Delta p$ signifies that the uncertainties in position and momentum are inversely related—the more accurately we know one, the less we can know the other.
- h is Planck's constant, a fundamental constant in quantum mechanics.
- The factor $\frac{1}{4\pi}$ adjusts the scale, symbolizing the limits of precision in measurements.

To Heisenberg, this was more than just a formula—it was a way of understanding the philosophical nature of reality. It suggests that at the quantum level, the universe is governed by probabilities, not certainties. There's a certain beauty in that, almost like art, where the unknown and the unpredictable become necessary parts of the creative process. Just as an artist works with light and shadow, a physicist works with certainty and uncertainty, revealing the deeper layers of nature's design.

At the moment of his breakthrough, Heisenberg was both exhilarated and unsettled by the implications of the uncertainty principle. As a physicist, he had long worked within the framework of classical determinism, where every cause had a predictable effect. To discover that nature itself imposed limits on what could be known felt almost like a betrayal of the scientific tradition. Yet, Heisenberg understood that this uncertainty was not a failure of science. It was an essential truth about the universe. In letters to his colleagues, Heisenberg expressed a mix of awe and hesitation, realizing that his work would upend centuries of scientific thinking.

The reaction from his peers was equally split. Some, like Einstein, famously resisted the idea, refusing to believe that God "played dice with the universe." Others recognized the effect of Heisenberg's discovery and quickly embraced it as a new way of seeing reality. His mentor, Niels Bohr, was incredibly supportive, seeing it as a natural extension of his ideas on quantum mechanics. Together, Heisenberg and Bohr paved the way for the next generation of physicists to explore the quantum world not with certainty but with curiosity and wonder. In the end, Heisenberg's uncertainty principle was more than a scientific milestone—it was a humbling reminder of humanity's limited grasp on the cosmos and a call to embrace the mysteries that remain unsolved.

Collaborations and Friendships

Heisenberg wasn't working in isolation during these transformational years. He was part of a community that included some of the greatest minds in physics. His friendship and teamwork with Wolfgang Pauli, Erwin Schrödinger, and Max Born played a vital role in developing quantum mechanics. Pauli, known for his sharp wit and no-nonsense approach, became one of Heisenberg's closest collaborators. The two regularly exchanged ideas, and their debates—often intense—helped shape the direction of quantum theory.

Heisenberg and Schrödinger's relationship was more complex. While they respected each other's work, their approaches to quantum mechanics were structurally different. Schrödinger's wave mechanics and Heisenberg's matrix mechanics initially seemed incompatible, leading to passionate rivalry. But, this feud eventually gave way to mutual understanding, and their combined insights formed the foundation of modern quantum mechanics. Heisenberg's ability to cooperate, even with those who had opposing views, was a testament to his openness and humility.

Heisenberg was also known for his approachable and warm personality, which made him a popular mentor for young scientists. He didn't see himself as someone separate from his students but rather as a fellow traveler on the same path of discovery. He would often invite younger physicists to his home for discussions that would stretch late into the night. These gatherings were informal, with Heisenberg encouraging everyone to speak freely and question everything. His ability to nurture an atmosphere of open dialogue made him a beloved teacher and mentor.

The Copenhagen Meeting

One of the most intriguing and controversial moments in Heisenberg's life was his 1941 meeting with Niels Bohr in Copenhagen. The meeting took place during World War II, at a time when Heisenberg was working on nuclear research in Germany, and Bohr had already fled to Sweden to escape Nazi persecution. The exact details of their conversation have remained a mystery, but it is widely believed that Heisenberg attempted to discuss the ethics of nuclear weapons with Bohr.

The meeting, visibly, did not go well. Bohr was reportedly shaken by the encounter, and their father-son relationship would be officially impossible to repair. Some historians believe that Heisenberg may have sought Bohr's guidance on how to prevent the development of nuclear weapons, while others suggested that Heisenberg was more concerned with protecting Germany's nuclear program. Whatever the truth, the meeting is a touching reminder of the ethical clashes that scientists may face, particularly in higher-stress situations, such as times of war.

Their meeting is often framed as a moral turning point for Heisenberg, a moment when he had to confront the potential consequences of his work. While the details may remain unclear, the emotional and philosophical weight of that encounter is undeniable. It adds a layer of complexity to Heisenberg's legacy, illustrating the difficulty of balancing progress with ethical responsibility.

His Post-War Life

After World War II, Heisenberg's life took a different turn. He was held in British custody for a time but was eventually released and returned to Germany, where he became a central figure in rebuilding the scientific community. He played a key role in establishing the Max Planck Institute for Physics and worked tirelessly to promote the peaceful use of nuclear energy. Heisenberg was determined that science should serve humanity, not destroy it, and he became an advocate for nuclear disarmament and international cooperation.

During this period, Heisenberg also returned to teaching, continuing to inspire a new generation of physicists. His post-war efforts to rehabilitate German science were marked by the same humility and intellectual curiosity that had defined his earlier work. He wasn't just rebuilding institutions—he was helping to restore trust in the scientific community.

Philosophy and Legacy

Throughout his life, Heisenberg maintained a deep interest in philosophy. He often reflected on the relationship between science and human knowledge, particularly in light of his uncertainty principle. For Heisenberg, this wasn't just a scientific concept; it was a philosophical one that had far-reaching implications for how we understand reality. He believed that science could never provide complete answers, only deeper questions.

In his later years, Heisenberg wrote broadly about the philosophical importance of quantum mechanics, drawing meaningful connections between physics and existence, knowledge, and ethics. His writings on these topics continue to be studied today, not just by physicists but also by philosophers and historians. Heisenberg's legacy is one of intellectual curiosity, humility, and resilience. He lived through some of the most explosive periods in modern history, yet his passion for science and his belief in the power of human knowledge never wavered.

Final Thoughts

Werner Heisenberg's life was a blend of music, philosophy, and science. His contributions to quantum mechanics changed the course of physics, but it was his ability to balance intellectual trials with personal reflection that made him stand out. Heisenberg wasn't just a man of equations and experiments—he was deeply introspective.

In many ways, Heisenberg saw the world through a harmonic lens, where science, much like music, required a sense of rhythm, balance, and uncertainty. This approach to life and science allowed him to embrace the unknown rather than fear it, and the courage to rethink what we believe is possible.

Erwin Schrödinger

Equations and Existentialism

Erwin Schrödinger is best known for his groundbreaking contributions to quantum mechanics and his famous thought experiment, Schrödinger's cat. Yet, the man behind mathematics was more than just a physicist—he was super clever with a genuine love for nature and philosophy. Schrödinger's scientific achievements not only reshaped physics, but his unique outlook on life, his partnerships with peers like Albert Einstein, and his eager curiosity across many fields make him a fascinating figure in the history of science.

Born in Vienna in 1887, Schrödinger grew up in a cultured and upscale household of well-educated parents. His father was a botanist, and his mother came from a family of respected academics. As a child, Schrödinger was naturally curious about the world around him, spending hours observing the beautiful details of nature. His upbringing was a balance between scientific fundamentals and philosophical inquiry that would define his life's work. He later recalled how his early exposure to the natural world and abstract debates helped him see science as more than a tool for discovery—it was a way to reflect on the substance of existence.

Schrödinger's Masterpiece

Schrödinger's most significant addition to quantum mechanics is his famous wave equation, formulated in 1926. The equation modernized how scientists understood particle behavior at the atomic level. Before Schrödinger, physicists like Niels Bohr and Werner Heisenberg had laid the groundwork for quantum theory, but it was Schrödinger's wave equation that gave the theory its mathematical backbone.

The Schrödinger equation describes how the quantum state of a physical system changes over time. In classical physics, the behavior of particles can be predicted with certainty, but Schrödinger's equation introduced a probability model. In this model, particles like electrons could exist in multiple states at once, and their exact location or momentum is described not by certainty but by a probability wave. It was a revolutionary concept that challenged traditional notions of determinism in physics.

But Schrödinger's equation wasn't just a tool for understanding particles—it also reflected his broader outlook on philosophy. Schrödinger believed that the universe wasn't governed by strict laws of cause and effect but by fluid and interconnected forces. For Schrödinger, the wave function wasn't just a mathematical abstraction. It was a reflection of the fundamental mystery of existence. In his later years, he often spoke about how science, philosophy, and nature were interconnected in ways that were still far beyond human comprehension.

A Thought Experiment that Endures

In 1935, Schrödinger introduced one of the most famous thought experiments in the history of science: Schrödinger's cat. The experiment, designed to highlight the contradiction of quantum mechanics, involves a cat placed in a sealed box with a radioactive atom, a Geiger counter, and a vial of poison. According to quantum mechanics, until the box is opened, the cat exists in a superposition of states—it is both alive and dead at the same time. Only when an observer opens the box does the cat's fate become certain.

Schrödinger imagined this thought experiment out of frustration. He was grappling with the strange implications of quantum theory—particularly the Copenhagen interpretation championed by his peers, Niels Bohr and Werner Heisenberg. For Schrödinger, the idea that a particle could exist in multiple states until observed seemed absurd when applied to everyday objects. To illustrate this, he created a situation that was so ridiculous, so extreme, that it would force people to confront the oddity in quantum mechanics. A cat, alive and dead at the same time? It was almost a joke, but a serious one, meant to challenge the dominant views in quantum theory.

His colleagues had mixed reactions. Niels Bohr and Werner Heisenberg, key advocates of the Copenhagen interpretation, found the thought experiment clever, though they didn't see it as a fatal critique. They viewed it more as a playful exercise rather than a serious rebuttal. Others, however, found it disturbing or even foolish. Albert Einstein, who also had reservations about quantum mechanics, sympathized with Schrödinger's explanation. In their letters, Einstein famously wrote about his distaste for the randomness implied by quantum mechanics, declaring, "God does not play dice with the universe." Schrödinger's cat, in a way, echoed Einstein's concerns—both men were uneasy about a world where chance and observation could shape reality.

Yet, despite the mixed reactions at the time, Schrödinger's cat has endured as one of the most memorable ways to explain the mysteries of quantum mechanics. Schrödinger himself was somewhat amused by the lasting legacy of the thought experiment. He once remarked that it was strange to be remembered for a hypothetical cat when his life's work had accomplished so much more. Though Schrödinger intended it as a playful criticism, the cat has taken on a life of its own, symbolizing the strange, often confusing world of quantum physics.

A Passion for Nature and Hiking

Outside the lab, Schrödinger had a deep love for nature and the outdoors. He was an avid hiker and often found solace in the mountains. He believed that time spent in nature offered a mental clarity that complemented his scientific work. To Schrödinger, the structure and harmony of the natural world mirrored the same balance in his equations.

His hikes through the Austrian Alps were often more than physical exercise—they were an intellectual retreat. He would take long walks, considering the mysteries of the universe, letting his mind roam as freely as the paths he followed. In letters to his colleagues, Schrödinger often wrote about how the beauty of nature inspired his work, particularly his explorations into quantum mechanics. He would sometimes pause mid-hike to jot down an equation or idea that had come to him while walking through the forests.

For Schrödinger, the connection between nature and science was inseparable. He believed that by studying the natural world, scientists could uncover the fundamental truths of existence. In his view, nature wasn't just a backdrop for scientific inquiry—it was a living, breathing part of the universe's grand equation.

Friendship and Collaboration

Schrödinger shared a special bond with Albert Einstein, one of the most influential figures in modern physics. The two men held each other in great mutual respect and often exchanged ideas about the philosophical substances of quantum mechanics. Though they approached physics differently—Einstein favoring a deterministic view while Schrödinger embraced quantum uncertainty—their debates fueled some of the most profound discussions in 20th-century science.

Schrödinger admired Einstein's courage and willingness to challenge the prevailing scientific norms of the time. The two regularly communicated, often discussing science, philosophy, and the nature of reality. In many ways, their friendship was defined by their shared interest in finding the deeper truths behind scientific equations. Both men believed that science could offer insights into the mysteries of existence, even if those insights were sometimes at odds.

In quantum mechanics, Schrödinger and Einstein often found themselves on the same side of debates, particularly in their criticism of the Copenhagen interpretation championed by Niels Bohr. Schrödinger's cat thought experiment was inspired by his discussions with Einstein, who famously disliked the probabilistic nature of quantum mechanics. Though they couldn't fully resolve their differences with the Copenhagen camp, Schrödinger and Einstein's friendship was a strong alliance that pushed the boundaries of scientific thought.

From Physics to Biology

Schrödinger was fiercely curious, and his interests extended beyond quantum mechanics. In the 1940s, he became increasingly interested in biology and the question of life itself. His influential book "What is Life?" was published in 1944. It aimed to bridge the gap between physics and biology by exploring how quantum mechanics could explain the fundamental processes of life.

In What is Life?, a book that I highly recommend, Schrödinger speculated about how genetic information might be stored in molecular structures, a concept that foreshadowed the discovery of DNA. While Schrödinger himself didn't discover the structure of DNA, his work inspired scientists like Francis Crick and James Watson, who later credited Schrödinger's ideas as a significant influence on their discovery of the double helix.

Schrödinger's foray into biology demonstrates his inherent curiosity and willingness to explore unknown regions. He saw that science had the capability to overlap, where advancements in one area could illuminate questions in another. His ability to easily move between fields made him one of the most versatile thinkers of his time.

Eastern Philosophy and Consciousness

Late in his life, Schrödinger became fascinated with Eastern philosophy, particularly the Hindu concept of Brahman, the idea of an underlying, universal consciousness. Schrödinger believed that the questions raised by quantum mechanics were not just scientific but deeply philosophical, touching on the nature of reality and human perception.

Though he was careful not to combine science and spirituality, Schrödinger saw parallels between quantum mechanics and the philosophical ideas of interconnectedness found in Eastern thought. He often reflected on how science and philosophy could work together to offer a more holistic understanding of the universe. For Schrödinger, the nature of consciousness and existence were just as worthy of exploration as the mysteries of the atom.

Collaborations, Rivalries, and Respect

Schrödinger was a central figure in the development of quantum mechanics, and his relationships with his fellow physicists were as dynamic and varied as his ideas. While his professional rivalry with Heisenberg and Bohr over quantum mechanics is well-documented, their interactions weren't solely defined by disagreement. Schrödinger, known for his philosophical leanings, often found himself at odds with Heisenberg's more mathematical approach, but the two maintained a mutual respect. In a memorable exchange during a conference,

Schrödinger joked, "I may not agree with your matrix mechanics, but I must admit, it's like listening to an intricate fugue—impossible to ignore." Intricate fugue = elaborate imagination

With Max Planck and Wolfgang Pauli, Schrödinger shared a collaborative spirit. In these discussions, Schrödinger thrived on exchanging ideas, bouncing between debates on wave mechanics and expansive reflections on existence. While their scientific differences were real, they were often framed within a broader, collegial respect for the pursuit of knowledge. Schrödinger's charm, combined with his intense curiosity, made him a beloved figure among his peers, someone who could challenge their theories yet share a glass of wine at the end of the day.

Schrödinger shared a special bond with Albert Einstein, one of the most influential figures in modern physics. The two men held each other in great mutual respect and often exchanged ideas about the philosophical substances of quantum mechanics. Though they approached physics differently—Einstein favoring a deterministic view while Schrödinger embraced quantum uncertainty—their debates fueled some of the most profound discussions in 20th-century science.

Schrödinger admired Einstein's courage and willingness to challenge the prevailing scientific norms of the time. The two regularly communicated, often discussing science, philosophy, and the nature of reality. In many ways, their friendship was defined by their shared interest in finding the deeper truths behind scientific equations. Both men believed that science could offer insights into the mysteries of existence, even if those insights were sometimes at odds.

Where Heisenberg and Bohr presented intellectual challenges, Einstein offered a philosophical ally—a kindred spirit in the search for the universe's deeper truths. Schrödinger and Einstein often found themselves on the same side of debates, particularly in their criticism of the Copenhagen interpretation championed by Niels Bohr. Schrödinger's cat thought experiment was inspired by his discussions with Einstein, who famously disliked the probabilistic nature of quantum mechanics. Though they couldn't fully resolve their differences with the Copenhagen camp, Schrödinger and Einstein's friendship was a strong alliance that pushed the boundaries of scientific thought.

Final Thoughts

Erwin Schrödinger's life was a fusion of science, philosophy, and a deep appreciation for the natural world. His versatility made him one of the most influential figures in modern science. His legacy endures not just in the equations that bear his name, but in how he approached the pursuit of knowledge. Science is sometimes about asking the right questions and embracing the unknown. For young scientists today, his life is an inspiring example of how to question established assumptions, engage with fields beyond their expertise, and apply principles from one discipline to solve complex problems in another.

PAUL DIRAC

THE SILENT INNOVATOR

Paul Dirac is widely regarded as one of the most brilliant theoretical physicists of the 20th century. Well known for his innovative contributions to quantum mechanics and the prediction of antimatter. Dirac's work forever changed modern physics. Yet, behind the scenes of his monumental achievements was a man of peculiar personality—a deeply introverted individual whose quiet demeanor hid a relentless passion for mathematics, nature, and aviation. His life and work offer us a glimpse into the mind of a scientific genius and a story of a man who struggled with human connection while forming strong academic bonds with peers like Einstein, Heisenberg, and Feynman.

Early Life and Foundation

Born in 1902 in Bristol, England, Paul Dirac grew up in an educator's household. His father, Charles Dirac, was a very strict Swiss teacher, while his mother, Florence Dirac, came from an English family. Dirac's early education was heavily influenced by his father's stern teaching methods, which included enforcing French as the only language spoken at home. This stifling environment contributed to Dirac's introverted nature, as he became more comfortable in silence than in conversation.

Despite this, Dirac excelled in academics, particularly in mathematics and physics, subjects that offered clear answers and avoided the uncertainty of language. Paul didn't like that words or expressions could have two or more possible ways of interpretation. He attended the University of Bristol, where he studied electrical engineering before switching to mathematics. His early education laid the groundwork for his later theoretical breakthroughs, providing him with a strong foundation in the mathematical style that would define his career.

Dirac's Guiding Philosophy

For Paul Dirac, equations weren't just tools for solving scientific problems—they were works of art. He wholeheartedly believed in the concept of "mathematical beauty," the idea that the more elegant a mathematical formulation was, the more likely it was to describe fundamental truths about the universe. This philosophy guided his work and set him apart from other physicists. Where some pursued practical solutions, Dirac sought mathematical purity.

This passion for elegant solutions was nowhere more clear than in the creation of the Dirac Equation. Formulated in 1928, the equation merged quantum mechanics and special relativity, providing the first complete description of the electron and its behavior at high velocities. Dirac's equation not only explained the existence of electron spin but also predicted the existence of antimatter. This revolutionary concept would be confirmed by Carl Anderson's discovery of the positron in 1932.

The Dirac Equation is often regarded as one of the most beautiful equations in physics, reflecting Dirac's belief that the aesthetic quality of mathematics could reveal deep truths about the natural world. His confidence in the elegance of his theory was such that when asked if he cared whether his work agreed with experiments, Dirac famously replied, "I am not interested in experimental results. I am interested in the truth."

The simplified form of the Dirac equation, often expressed in matrix notation, is:

$$\left(i\gamma^0\frac{\partial}{\partial t} + i\gamma^1\frac{\partial}{\partial x} + i\gamma^2\frac{\partial}{\partial y} + i\gamma^3\frac{\partial}{\partial z} - mc\right)\psi = 0$$

In this equation:

- **Gamma Matrices ($\gamma^0, \gamma^1, \gamma^2, \gamma^3$):** These represent a set of 4x4 matrices commonly used in quantum field theory to account for spin in relativistic particles.
- **Wavefunction (ψ):** This symbol represents the wavefunction, describing the quantum state of the particle.
- **Mass (m):** Denotes the mass of the particle.
- **Speed of Light (c):** The constant speed of light in a vacuum.
- **Imaginary Unit (i):** This is the imaginary unit, essential in complex calculations in quantum mechanics.
- **Partial Derivatives ($\partial/\partial t$, $\partial/\partial x$, etc.):** These show how the wavefunction changes with respect to time and space.

Introversion and Silence

Dirac's colleagues often described him as the "silent genius." He was notorious for his limited speech, often replying to questions with only a word or two. This peculiar trait led to jokes about "Dirac units," with one Dirac unit equaling one word per hour of conversation. Yet, while some incorrectly saw his shyness as disdain, it was, in fact, an essential part of how he approached his work. Silence gave Dirac the space to think deeply, allowing him to study the complexities of theoretical physics without distraction.

Despite his introverted nature, Dirac formed meaningful relationships with other leading physicists of his time. His friendship with Albert Einstein was one of mutual respect. Although they approached physics differently—Einstein was more inclined toward deterministic views, while Dirac embraced quantum probability—their discussions about the deeper truths of the universe were wise. Dirac admired Einstein's resistance to the idea of a probabilistic universe but found a way to reconcile quantum mechanics with relativity through his own groundbreaking work.

Dirac's extreme quietness did not prevent him from forming other key friendships, either. His relationship with Werner Heisenberg was significant in advancing quantum mechanics. Heisenberg's matrix mechanics influenced Dirac's thinking, and the two worked closely to refine the early formulations of quantum theory. Even with their differing approaches, Heisenberg respected Dirac's silent intensity and unique insights, seeing him as a vital contributor to the emerging quantum field.

The Prediction of Antimatter

Perhaps the greatest of Dirac's predictions was the existence of antimatter. His Dirac Equation not only described the electron but also predicted the existence of a counterpart—an identical particle with an opposite charge, known today as the positron. This prediction was radical at the time, as no one had ever observed such a particle. Many physicists were skeptical, but Dirac remained confident in the beauty and logic of his equation.

When Carl Anderson discovered the positron in 1932, Dirac's theoretical prediction was vindicated, establishing his place as one of the great minds of modern physics. When Dirac realized that his equation predicted the existence of antimatter, he felt a quiet sense of awe rather than elation. It wasn't triumph he sought, but rather a deeper understanding of the universe's hidden symmetries—a reflection of his lifelong belief in the beauty of mathematics. The discovery of antimatter opened new avenues of research, contributing to advancements in particle physics, cosmology, and even medical imaging technologies like PET scans.

This discovery shows Dirac's ability to bridge the gap between theory and reality. His prediction of antimatter, born entirely from mathematical reasoning, showed the power of pure thought in driving discovery. It was a testament to Dirac's belief that mathematical beauty could reveal aspects of the universe that had yet to be seen.

Collaborations, Rivalries, and Friendships

Though Dirac was often socially isolated, he formed key partnerships with other physicists, which were essential to his scientific progress. His collaboration with Enrico Fermi led to the development of Fermi-Dirac statistics, a fundamental tool for understanding the behavior of fermions in systems at very low temperatures. Fermi, known for his more outgoing personality, worked well with Dirac, complementing his silent intensity with a more conversational approach to problem-solving. They were an unlikely duo but balanced each other perfectly.

Dirac's relationship with Niels Bohr, however, was more complex. Bohr, known for his bustling institute in Copenhagen and his love of debates, often clashed with Dirac's quiet, solitary style. Yet, despite their different personalities, they shared a deep intellectual respect. Bohr valued Dirac's contributions to quantum mechanics, even if their approaches to understanding the theory differed. Dirac's solitary walks in Copenhagen—where he silently worked through the equations in his mind—contrasted sharply with Bohr's lively, discussion-filled gatherings. Still, both men recognized the value in each other's methods.

One of Dirac's most important intellectual friendships was with the eccentric physicist Richard Feynman. Feynman admired Dirac's precision and often referenced Dirac's work as a guiding influence on his own research. Despite their vastly different personalities—Feynman being known for his showmanship and humor—the two shared a deep bond, rooted in their mutual love for theoretical physics.

Passions Outside of Physics

Despite his intense focus on physics, Dirac found time to nurture other passions, particularly his love for aviation and hiking. While Dirac never became a pilot, he was fascinated by the mechanics of flight and often followed developments in aviation with great interest. His interest in the movement of machines through space mirrored his scientific investigations into the motion of particles and forces in the cosmos.

Hiking was another passion for Dirac, providing him with the solitude he needed to process complex mathematical problems. He often took long walks through the English countryside or the mountains, finding peace in nature's quiet beauty. These solitary hikes became essential to his thinking process, offering him the mental clarity to tackle the most challenging theoretical problems.

Dirac's personal life, while quiet, was anchored by his marriage to Margit Wigner, the sister of physicist Eugene Wigner. Margit provided Dirac with emotional stability and support, helping him balance the demands of his work with a peaceful home life. Their marriage was a partnership in the truest sense, allowing Dirac to maintain the solitude he needed for his scientific work while also offering him a connection to the outside world.

Final Thoughts

Paul Dirac was a man of few words but immense depth. He bridged the gap between quantum mechanics and relativity, predicted the existence of antimatter, and influenced generations of physicists. Paul Dirac remains one of the most influential figures in physics. Though he struggled with social interactions, Dirac still formed meaningful partnerships with key figures like Heisenberg, Fermi, and Einstein, and his friendship with Richard Feynman continued to inspire new ideas in theoretical physics.

The legacy of Dirac's work continues to resonate throughout modern science. His groundbreaking ideas have influenced everything from cosmology to medical imaging, proving that theoretical advancements can have practical applications far beyond their initial scope. For young scientists, Dirac's life offers a powerful lesson: that brilliance and innovation don't always come from the loudest voice in the room but maybe from the quiet pursuit of truth, driven by a passion for discovery and a belief in the beauty of the universe's underlying structure. It shows that innovation may come from stepping away from the crowd, embracing solitude, and allowing the mind to wander into uncharted territory. His work teaches us that the courage to trust in the elegance of your ideas, even when the world has yet to understand them, can lead to discoveries that reshape the future.

RICHARD FEYNMAN

MASTER OF SIMPLICITY

Richard Feynman was more than just a superstar physicist; he was an educator, a philosopher of science, and an endlessly curious human being. Known for his playful approach to learning and his unique ability to break down complex ideas, Feynman transformed how quantum mechanics was taught and understood. His contributions not only reshaped modern physics but also left a lasting legacy in education, where his emphasis on simplicity, curiosity, and directness continues to inspire students and scientists alike. We will now explore Feynman's unique blend of brilliance, humor, and hobbies that made him a truly special figure in the history of science.

Feynman Diagrams: Simplifying the Complex

Feynman's most significant contribution to quantum mechanics is undoubtedly the Feynman diagrams. These visual tools, introduced in 1948, revolutionized the way physicists understood particle interactions. Before Feynman diagrams, the calculations involved in quantum field theory were complex, requiring dense algebraic manipulations. Feynman's diagrams offered a visual representation that made these interactions intuitive and accessible.

The diagrams depict particles as lines and their interactions as vertices, allowing scientists to "see" the process rather than fumbling through complex equations. This simplified the math and opened quantum mechanics to a broader audience. For Feynman, the diagrams embodied his belief that "If you can't explain it simply, you don't understand it well enough." His talent was to make the abstract more tangible, giving students and fellow scientists a tool to cut through complexity.

But Feynman didn't just stop at creating the diagrams—he made them a cornerstone of his teaching. In lectures at Caltech and beyond, Feynman used the diagrams to illustrate the behavior of particles in a way that was engaging, direct, and often fun. His ability to break down the most complex scientific concepts into simple, relatable explanations is part of what made him such a beloved teacher.

A Playful Approach to Science and Life

One of Feynman's most endearing qualities was his sense of play. He approached science the way a child might approach a puzzle: with wide-eyed curiosity and an eagerness to explore. This playfulness extended beyond his work as a physicist. Feynman famously loved practical jokes, and during his time at Los Alamos, while working on the Manhattan Project, he developed a habit of cracking safes containing top-secret information. What started as a bit of fun became a minor obsession, as Feynman sought to demonstrate how important it was to take security seriously, even if through unconventional means.

His playfulness wasn't limited to pranks. Feynman had a passion for music, particularly the bongos. He often played them at parties, and they became a central part of his life. In fact, Feynman's love for drumming was so profound that he sometimes performed with samba groups while visiting Brazil. He believed that music, like physics, had its own structure and beauty that could be understood and appreciated deeply, even if approached playfully.

Feynman's colleagues often commented on how his sense of wonder fueled his creativity. Whether investigating how ants navigate, the physics of spinning plates, or the properties of water in a sprinkler, Feynman's curiosity knew no bounds. He embodied the idea that science is not just a career or a subject but a way of looking at the world—a lens through which everything can be explored.

Relationships with Fellow Scientists

Feynman's relationships with his peers were as varied and dynamic as his interests. He worked closely with some of the most influential physicists of his time, including Niels Bohr, Murray Gell-Mann, and Julian Schwinger. While he deeply respected their contributions, Feynman often approached these relationships with a characteristic mix of humility and challenge.

His relationship with Bohr, for instance, was one of mutual respect, but it wasn't without disagreements. Bohr, a towering figure in quantum mechanics, was known for his long, complex explanations, which often left Feynman frustrated. "I never understood what he was saying," Feynman once remarked. But even in confusion, Feynman found inspiration. While difficult to grasp, Bohr's ideas pushed Feynman to think in new ways.

With Murray Gell-Mann, Feynman's relationship was more competitive. The two worked side by side at Caltech, but they had vastly different styles. Gell-Mann, known for his meticulousness, contrasted sharply with Feynman's more relaxed, intuitive approach to science. Despite these differences, their rivalry produced some of the most important breakthroughs in particle physics during the 20th century.

But it wasn't just about competition—Feynman also formed deep friendships with fellow physicists, particularly with Nobel Prize winners like Julian Schwinger and Richard Feynman's student, Freeman Dyson. Their teamwork, often punctuated by jokes and laughter, produced some of the era's most significant scientific insights, including contributions to quantum electrodynamics (QED).

In the Face of Loss

While Feynman was known for his humor and lightheartedness, he also faced profound personal challenges. One of the defining moments of his life was the death of his first wife, Arlene. Feynman had married Arlene during the early days of World War II, regardless that she was gravely ill with tuberculosis. Her death in 1945 was a devastating blow, but it also revealed a different side of Feynman—the side that dealt with grief not by shutting down but by finding meaning in life's hardships.

Feynman's coping mechanism was not to hide his grief but to be completely open and continue living fully, even amidst sorrow. He wrote heartfelt letters to Arlene after her death, expressing his undying love and yearning for her presence. These letters, discovered after his death, revealed a vulnerability that contrasted with his public persona as the carefree, witty scientist.

This period of Feynman's life shifted his approach to science and living. He became more focused on enjoying the present and life's uncertainties. His bongo playing, his adventures in art, and his willingness to try new things all stemmed from this profound loss. Feynman knew, perhaps better than anyone, that life was fragile and fleeting, and this realization fueled his passion for discovery and joy in all things.

The Challenger Disaster

One of the most remarkable chapters of Feynman's later career was his involvement in the investigation of the 1986 Challenger Space Shuttle disaster. Appointed to the Rogers Commission, Feynman's direct and straightforward approach quickly became a key factor in uncovering the technical cause of the tragedy.

While the investigation was bogged down in bureaucratic discussions, Feynman famously demonstrated the cause of the disaster—O-ring failure—using nothing more than a glass of ice water. His simple experiment showed how the rubber O-rings used in the shuttle's booster rockets became brittle in cold temperatures, leading to the catastrophic failure. This practical, hands-on approach was symbolic of Feynman's belief that even the most complex problems could be understood through simple, direct methods.

Feynman's role in the investigation saved NASA from future tragedies and demonstrated the value of clear thinking and skepticism. He was critical of the bureaucratic culture at NASA, which had allowed safety concerns to be ignored. While not always appreciated by the administration, Feynman's bluntness was instrumental in preventing future disasters.

Teaching and Communication

Feynman's influence on science education cannot be overstated. His famous "Feynman Lectures on Physics" remains one of the world's most widely-read series of physics textbooks. These lectures are still celebrated for their ability to make advanced topics accessible and engaging for students of all levels.

Feynman's unique teaching style—characterized by humor, enthusiasm, and a refusal to take anything too seriously—left a lasting impact on generations of students. He believed that learning should be fun and that science, at its heart, was about curiosity and discovery. His legacy lives on in the way science is taught today, with many educators adopting his approach of using real-world examples and visual aids to make learning more engaging.

But Feynman's contributions to science communication went beyond the classroom. He was a master storyteller, able to take complex scientific principles and turn them into narratives that anyone could understand. His books, like "Surely You're Joking, Mr. Feynman!" and "The Pleasure of Finding Things Out," are still beloved by readers for their humor, wit, and insight into the mind of one of the greatest scientists of the 20th century.

Final Thoughts

Richard Feynman was a man of contradictions: playful yet profound, introverted yet charismatic, deeply intellectual yet refreshingly satirical. His contributions to quantum mechanics left an iconic mark on the world. But perhaps more importantly, Feynman's life was a testament to the power of curiosity, the joy of discovery, and the importance of living fully—even in the face of uncertainty.

For young scientists today, Feynman's life offers a valuable lesson: that true understanding doesn't come from complexity but from clarity and that science is best approached with curiosity and an open mind. His legacy teaches us that impressive discoveries often come from asking simple questions and finding joy in the process of exploration. Whether through diagrams, storytelling, or hands-on demonstrations, Feynman showed that science, at its core, is a journey that thrives on simplicity, honesty, and a genuine love for learning.

JOHN BELL

BEYOND INEQUALITY

John Bell was more than just a brilliant physicist; he was a thinker deeply engaged in quantum mechanics. His groundbreaking work on Bell's Theorem reshaped how we understand the universe, challenging long-held views about the nature of reality. But Bell's contributions extended beyond science. His personal relationships, curiosity, and love for philosophy made him a remarkable figure in 20th-century physics. Let's dive into Bell's scientific achievements, his personal side, and the fantastic thoughts of a man who forever altered the course of quantum mechanics.

Engineering Roots and Scientific Rigor

Prior to his involvement in quantum mechanics, John Bell sharpened his skills as an engineer." Specifically, he studied electrical engineering at Queen's University Belfast, an experience that significantly shaped his approach to science. The built-in process of engineering made Bell a methodical and systematic thinker, traits that would become complimentary in his later work on quantum theory.

Bell's engineering mindset gave him a unique ability when he turned to theoretical physics. In a field often dominated by abstract thought, Bell brought a problem-solving mentality that focused on finding tangible, workable solutions. The strict engineering rules helped him approach unique problems with a certain level of humility. He wasn't interested in chasing fame; instead, he focused on solving real-world puzzles. This down-to-earth nature made him admired by his colleagues and students, many of whom found Bell's calm approach to scientific inquiry refreshing.

In fact, Bell's painstaking attention to detail would be essential in his later work, particularly in crafting Bell's Theorem. This Theorem would challenge the core beliefs of classical physics and demand the same rigor and precision Bell had mastered during his engineering work.

Shattering Classical Assumptions

The core of John Bell's scientific legacy is Bell's Theorem, a principle that challenges classical views of physics. Bell's Theorem mathematically proves that no theory based on local hidden variables could reproduce all the predictions in the quantum world. At the heart of this discovery lies the phenomenon of quantum entanglement—particles that, once entangled, remain connected across vast distances. One particle's state instantaneously affects the other's state, regardless of the distance separating them. Albert Einstein famously dismissed this idea as "spooky action at a distance," but Bell showed that this "spookiness" was real.

Bell's Theorem had profound implications. It directly challenged the classical concept of locality—the idea that objects are only affected by their immediate surroundings. Instead, Bell's work suggested that quantum mechanics allows for non-local interactions, where changes in one part of the universe could instantly affect another, no matter the distance. This discovery forced physicists to reconsider the boundaries of space and time, prompting deep philosophical questions about the nature of reality itself.

When Bell published his Theorem in 1964, it sent shockwaves through the scientific community. His work demonstrated with mathematical certainty that the classical view of a deterministic, local reality was incompatible with quantum mechanics. For decades, physicists had been trying to grasp the strange behavior of quantum particles with a classical understanding of the world, but Bell showed that was impossible. This opened the door to new interpretations of quantum mechanics and laid the groundwork to experiment with quantum entanglement.

John Bell's reaction to his work was notably quiet, in keeping with his calm and reflective personality. Bell himself didn't describe his discovery in terms of emotions but rather as a necessity. It was a way to resolve a long-standing issue in the interpretation of quantum mechanics. His feelings leaned more toward curiosity and determination to explore the implications rather than chasing a feeling. Bell's Theorem was part of his personal quest to understand whether hidden variables could explain quantum phenomena in a way that fit within a realist and local framework. When his work led to the conclusion that no local hidden variable theory could replicate the predictions of quantum mechanics, Bell was definitely intrigued by the outstanding implications, but he wasn't one to display dramatic emotions about it.

As for his peers, the initial response wasn't too positive. In fact, Bell's Theorem initially received a somewhat lukewarm reception from the broader scientific community. Many physicists at the time were content with the Copenhagen interpretation of quantum mechanics, which focused on probability and the role of observation. Some physicists, like Niels Bohr and Werner Heisenberg, were uncomfortable with hidden variables from the start, and Bell's work seemed to challenge their ideas by raising questions about the nature of reality that many were hesitant to address.

Bell's results didn't immediately spark widespread interest or change the course of research overnight. His Theorem was subtle and highly mathematical, and while some physicists recognized its importance, it took time for the full significance to sink in. In fact, it wasn't until the 1970s and 1980s—after Alain Aspect conducted key experiments that tested Bell's inequalities—that the importance of Bell's work became widely accepted. Aspect's experiments confirmed that quantum mechanics does indeed violate Bell's inequality, meaning that the universe behaves in ways that defy local hidden variables.

Show Your Work

The key mathematical insight of Bell's Theorem is encapsulated not by a single equation but by something called Bell's inequalities. These inequalities are designed to test whether the results of quantum experiments can be explained by local hidden variables or if they require the non-local interactions predicted by quantum mechanics.

One of the simplest forms of Bell's inequality is the CHSH inequality (after Clauser, Horne, Shimony, and Holt, who reformulated Bell's work). The CHSH inequality is often used in experiments that test Bell's Theorem.

The inequality is typically written as:

$$\left|E(a,b) + E(a,b') + E(a',b) - E(a',b')\right| \leq 2$$

Here's what it means:

- E(a, b) represents the correlation between measurements made on two entangled particles using different settings a and b (e.g., angles or other settings on the measuring device).
- The terms E(a, b'), E(a', b), and E(a', b') represent correlations measured with other combinations of settings.

This CHSH inequality is the most commonly tested form of Bell's inequality in experiments. It's a powerful way to show the mathematical backbone of Bell's Theorem without getting too deep into abstract math.

Relationships and Intellectual Bonds

John Bell was not just an isolated genius working alone in a back office—his personal relationships played a big role in shaping his ideas. One of those relationships in Bell's life was with his wife, Mary Ross Bell, who was a physicist in her own right. Their marriage was more than a personal partnership but also an intellectual alliance. Mary often acted as a sounding board for Bell's ideas, providing him emotional support and scientific wisdom. Together, they shared a deep commitment to understanding the fundamental questions of quantum mechanics.

This partnership between John and Mary was particularly special. Bell often credited Mary for helping him stay grounded, and her mark was visible in many aspects of his life. Their collaboration reminds us that personal relationships and shared ideas often influence even the most complex scientific breakthroughs.

Bell's professional relationships were equally influential. He admired David Bohm, a physicist who challenged the mainstream Copenhagen interpretation of quantum mechanics. Bohm's work on hidden variables provided an alternative view of quantum theory that resonated with Bell's realist inclinations. Bohm's willingness to go against the grain inspired Bell, giving him the courage to question accepted norms and explore alternative explanations for quantum phenomena. This connection with Bohm shaped much of Bell's quantum concepts.

However, not all of Bell's relationships were pleasant. His interactions with physicists like Niels Bohr and Werner Heisenberg were often contentious. Bell was uncomfortable with their insistence that the observer played a fundamental role in shaping reality, as it conflicted with his belief in an objective, observer-independent world. Despite these differences, Bell respected his colleagues, even when they clashed on the interpretation of quantum mechanics. These debates, though heated, pushed Bell to sharpen his ideas, forcing him to refine his Theorem and its effects.

A Radical Shift

One of the most extreme ideas to emerge from Bell's Theorem is the concept of non-locality. This idea—that entangled particles can interact instantaneously across vast distances—defies the classical notion of locality and raises profound questions about the very fabric of space-time. How could particles seemingly communicate faster than the speed of light, violating the principles of Einstein's relativity?

For Bell, non-locality wasn't just a theoretical curiosity but a pathway to understanding quantum mechanics on a deeper level. His Theorem paved the way for experimental proof by physicists like Alain Aspect, whose groundbreaking experiments in the 1980s provided empirical evidence for Bell's predictions. Aspect's tests demonstrated that entangled particles behaved exactly as Bell had predicted, with interconnections that could not be explained by local hidden variables. These experiments were a turning point in the acceptance of quantum entanglement, solidifying Bell's place as one of the most significant figures in 20th-century physics.

Bell's work on non-locality didn't just transform quantum theory—it also opened the door to practical applications. Quantum cryptography, which relies on the behavior of entangled particles to ensure secure communication, is one of the most direct outcomes of Bell's work. Similarly, quantum computing, with its use of qubits in superposition and entanglement, draws heavily on the principles of non-locality that Bell helped establish. To learn more about the science, technology, and impact of quantum computing, please see my best-selling book titled Quantum Computing Explained for Beginners.

Hobbies and Personal Interests

Outside of his work, Bell was a man of many interests. He deeply loved classical music and enjoyed composers like Bach, Beethoven, and Mozart. These composers are known for their intricate compositions, much like the mathematical structures Bell worked on in his research. Bell appreciated the precision and harmony in their works, seeing parallels between the beauty of music and the elegance of physics. Classical music's structured yet expressive nature often provided Bell with a peaceful environment for deep thinking.

Music wasn't limited to enhancing his work environment. It played a broader role in his life, serving as a source of personal enjoyment and a way to unwind. Bell used music to create a warm, reflective atmosphere. Music was a backdrop to his life, helping him connect with the emotional depth that often contrasted with his calm and logical demeanor. Whether relaxing with a favorite symphony or playing music during dinner or quiet evenings with his wife Mary, Bell found that music enriched his work and home. Creating a space where he could reflect, recharge, and reconnect with the people and ideas that mattered most.

Bell was also known for his love of long, thoughtful walks. Much like Albert Einstein, Bell found that walking helped him think more clearly. These walks gave him the time and space to process ideas, often leading to new insights. His peers frequently noted how approachable and thoughtful Bell was, a man who remained humble regardless of his impressive list of achievements.

John Bell often took long, pondering walks near his home, particularly in Geneva, where he worked at CERN. The serene Swiss countryside provided the perfect backdrop for his relaxed nature. Bell's walks would sometimes take him along the banks of Lake Geneva or through the lush green fields and hillsides surrounding the area. These peaceful, scenic routes allowed Bell to escape the pressures of his scientific work and let his mind wander freely, often resulting in new breakthroughs or improvements to his theories.

His walks were not short strolls—they could last for hours, especially on weekends when he had more time. Bell was known to walk alone, enjoying the solitude as an additional way to recharge and think deeply about his work. Though his destinations were rarely grand or symbolic, the Swiss countryside's quiet landscapes and natural beauty became a source of inspiration and clarity. His routine hikes helped him balance the intense mental demands of his profession and the simplicity of being in nature, a habit that many of his colleagues admired.

Final Thoughts

John Bell straddled the line between philosophy and science, between the theoretical and the practical. His work on Bell's Theorem shattered classical notions of reality and opened the door to some of the most significant advancements in modern physics, from quantum cryptography to quantum computing. But beyond his groundbreaking contributions to science, Bell's life was a testament to curiosity, intellectual humility, and teamwork.

For young scientists, Bell's journey offers a valuable lesson: science is not just about proving theories but about asking the right questions. Even when faced with uncomfortable truths, Bell remained committed to exploring new ideas with openness and strict proof. His legacy shows that pursuing knowledge demands curiosity, perseverance, and the support of others in facing science's toughest challenges.

SATYENDRA NATH BOSE

A PARTNERSHIP WITH EINSTEIN

Satyendra Nath Bose's partnership with Albert Einstein represents a memorable chapter in the history of physics. Together, they reshaped how we understand quantum mechanics. Their collaboration offered groundbreaking insights into how particles behave at the quantum level, challenging existing theories and opening doors to future inventions. Beyond the technical achievements, we will explore a story of creativity, perseverance, and the meeting of different cultural worlds. We'll also take a closer look at Bose's hobbies, his influence beyond quantum mechanics, and how his life remains an inspiration for scientists today.

Ahead of His Time

Bose was an innovative thinker, willing to question established norms in his pursuit of knowledge. Born in 1894 in India, he grew up during a time when Western values heavily influenced education and exploration. Bose was part of the Bengal Renaissance, a cultural period in history marked by a revival of classical learning and wisdom. This movement blossomed in the late 19th and early 20th centuries and played a crucial role in shaping Satyendra Nath Bose's scientific exploration. During this period, India experienced a resurgence of interest in philosophy, arts, and science, blending Eastern and Western thought in a unique way.

This movement fostered a deep appreciation for education, self-reliance, and intellectual freedom, values that resonated deeply with Bose. Growing up in this environment, Bose was encouraged to think critically and question established norms, which later defined his approach to physics.

His early exposure to Western scientific methods, paired with his philosophical Indian heritage, gave him the tools to tackle complex problems in the quantum domain.

Bose's participation in the Bengal Renaissance wasn't merely passive; he embodied the spirit of the movement. Driven by a desire to contribute to the global scientific community, he pursued his studies at Presidency College in Calcutta, where he excelled in mathematics and physics. Bose was deeply inspired by thinkers like Rabindranath Tagore, who advocated for the fusion of science and spirituality. This worldview helped Bose see physics not just as a set of equations but as a pursuit that could reveal deeper truths about the universe. This transformation from the Bengal Renaissance allowed Bose to approach quantum mechanics with a fresh perspective, enabling him to challenge conventional Western scientific thinking.

One of Bose's most remarkable qualities was his ability to think outside the box. His work focused on indistinguishable particles—those that cannot be differentiated from one another. Bose's method defied the established views of the time, as he proposed that these particles could simultaneously occupy the same quantum state, an idea that classical physics could not explain. This innovative idea laid the foundation for what would become Bose-Einstein statistics, which greatly impacted the understanding of quantum particles.

Example of a Simple Equation (without getting too complex):

One of the key ideas behind Bose-Einstein statistics is that the distribution of bosons in different energy levels follows a specific formula:

$$n_i = \frac{1}{e^{(\epsilon_i - \mu)/kT} - 1}$$

Where:

- $\mathbf{n_i}$ represents the number of particles in the energy level $\boldsymbol{\varepsilon_i}$.
- $\boldsymbol{\mu}$ is the chemical potential, essentially how much energy it "costs" to add a particle.
- **k** is the Boltzmann constant, a fundamental constant that relates temperature to energy.
- **T** stands for the temperature of the system.
- **e** is the exponential function, used to model the distribution of particles.

Persistence in the Face of Rejection

Bose's journey had its challenges. Despite his innovative ideas, Bose faced significant hurdles in getting his work recognized. European journals initially rejected his 1924 paper on quantum statistics, a setback that would have discouraged many. However, Bose's resilience was one of his defining traits. Rather than giving up, he reached out directly to Albert Einstein, sending his paper to the influential physicist in the hopes of gaining his support.

Einstein immediately recognized the significance of Bose's work. In a gesture that would change the course of quantum mechanics, Einstein translated Bose's paper into German and helped secure its publication in the prestigious journal Zeitschrift für Physik. Einstein validated Bose's theories and expanded on them, applying Bose's principles to atoms and solidifying the foundations of what would later be called Bose-Einstein statistics.

This collaboration demonstrated the importance of perseverance and the power of partnerships. Bose and Einstein's teamwork transcended geographical and cultural boundaries, highlighting the universal nature of scientific inquiry. For Bose, the challenges he faced in getting his work published only strengthened his determination, and his fateful meeting with Einstein became a testament to the value of pushing forward even in the face of adversity.

Bose-Einstein Condensates

Bose and Einstein's collaboration led to the theoretical prediction of Bose-Einstein condensates (BECs), a state of matter that occurs at temperatures just above absolute zero. In this state, particles lose their individual identities and merge into a single quantum body, behaving as one unified system rather than as separate particles. This phenomenon refuses to obey the typical behaviors observed in everyday materials and shows the quirky nature of quantum mechanics.

While Bose and Einstein could only predict the existence of BECs in theory, their efforts laid the groundwork for decades of experimental research. It wasn't until 1995 that BECs were tested in the lab by physicists Eric Cornell and Carl Wieman, who created a condensate using rubidium atoms. This experiment confirmed the theoretical predictions made by Bose and Einstein more than 70 years earlier and opened the door to new avenues of research in quantum mechanics.

Today, BECs are used in a wide range of cutting-edge technologies. For instance, BECs have been employed to slow down light, providing new opportunities for research in quantum computing and precision measurements. Bose and Einstein's work is a shining example of how theoretical insights can lead to future discoveries that transform fields like condensed matter physics and statistical mechanics.

Building Partnerships & Respect

The partnership between Bose and Einstein was more than just a meeting of minds; it was a friendship rooted in mutual respect and academic pursuits. Bose's fresh, innovative ideas caught the attention of Einstein, whose open-mindedness allowed him to recognize the value of a new approach. Together, they formed a powerful bond that spurred some of the most important discoveries in modern physics.

What makes their collaboration particularly fascinating is the cultural thought diversity that enriched their work. Bose came from India, a land drowning in censorship but also rich in heritage. Bose's humility and Einstein's free thinking led to new perspectives essential to all their success. Their partnership set a precedent for international scientific cooperation, paving the way for future generations of scientists to work together across cultural and geographic divides.

One of the fascinating aspects of Bose's partnership with Einstein was how they navigated the challenges of communication. English was not Bose's first language, yet he managed to articulate complex ideas with such clarity that Einstein immediately grasped their significance. Their collaboration demonstrated that communication is possible in scientific discourse, even when working across cultural and linguistic boundaries.

Despite potential language barriers, Bose's ability to convey his groundbreaking ideas effectively serves as an inspiring message for young scientists. It shows that determination and expressing your thoughts can overcome even the most challenging obstacles, and that great ideas can transcend language and cultural differences.

Bose's Interests

Beyond his work in quantum mechanics, Bose was a man of many interests. He believed that science was interconnected and that knowledge in one field could enhance understanding in another. Bose's curiosity extended beyond physics—he had a passion for chemistry, mathematics, and philosophy. He often approached scientific problems holistically, blending insights from different disciplines to find innovative solutions.

One of Bose's most surprising passions was music, mainly Indian classical music. He found that music provided a creative outlet that complemented his scientific work. He was a keen listener and a talented musician in his own right. Bose often spoke of the parallels between music and science, seeing both as expressions of beauty and harmony. Music helped him relax and think more clearly, offering a way to balance the intense demands of his work life.

While Satyendra Nath Bose is best known for his contributions to quantum mechanics, his influence extended far beyond this one area of physics. Bose was deeply invested in the advancement of science in India and dedicated much of his life to teaching and mentoring young learners. After returning from Europe, Bose became a professor and administrator at various institutions, including the University of Dhaka and later the University of Calcutta, where he worked to elevate the quality of science education. Bose saw teaching as a perfect outlet for his mission to empower future generations of physicists and mathematicians, believing that science could serve as a tool for national and individual liberation from the mental confines imposed by the old guard.

I want to discuss his love for literature, particularly the works of Bengali writers like Rabindranath Tagore, whose poetic and philosophical approach deeply resonated with Bose's hunger for connection. Tagore's exploration of human emotions and societal themes influenced Bose's worldview, and he admired how literature could express complex ideas about life and existence, much like the abstract principles of quantum mechanics. Bose often enjoyed reading in quiet moments, finding solace and inspiration in poetry and philosophical essays. He sometimes shared his favorite pieces with colleagues and students, encouraging them to think broadly and see connections between science, philosophy, and the arts.

This multidimensional approach made Bose a holistic thinker, blending the rational with the creative in both his work and personal life. His influence went beyond the technical details of his scientific discoveries, encouraging a generation of students to break free from narrow academic boundaries and explore new, creative avenues of thought. Bose's humbleness and dedication to teaching left a lasting impact on his students, many of whom became very influential.

A Lasting Impact

The teamwork between Bose and Einstein advanced our understanding of quantum mechanics and paved the way for significant technological innovations. The theoretical foundation they laid for Bose-Einstein statistics and condensates has had lasting impacts on fields such as quantum computing, atomic clocks, and quantum sensors. These technologies are critical to advancements in communication, data processing, and precision measurement and continue to shape the future of science and industry.

Final Thoughts

Satyendra Nath Bose's life and work are a powerful reminder of the importance of curiosity, determination, and collaboration. His partnership with Albert Einstein challenged long-held views in physics, leading to some of the most important discoveries in modern science. For young scientists today, Bose's journey underscores the idea that innovation often comes from thinking outside the box and embracing free speech. His legacy reminds us that scientific progress depends not only on individual brilliance but also on the ability to collaborate, communicate, and stay in the pursuit of knowledge.

DAVID BOHM

UNVEILING HIDDEN VARIABLES

David Bohm's work on quantum mechanics and philosophy represents one of the most intriguing and challenging academic paths of the 20th century. Known for his groundbreaking ideas on quantum theory, particularly his development of Bohmian mechanics or hidden variables theory, Bohm dared to challenge the prevailing Copenhagen interpretation. Bohm was a thinker who bridged the gap between science and philosophy, expanding into consciousness, dialogue, and the very nature of reality. His contributions, along with his struggles against political abuse and professional isolation, provide modern scientists with a compelling narrative of resilience, courage, and the pursuit of knowledge.

Challenging the Status Quo

David Bohm's creation of Bohmian Mechanics was closely linked to his personal views about the nature of reality. Unlike the prevailing view in physics, which accepted uncertainty as a fundamental aspect of quantum mechanics. Bohm was uncomfortable with the notion of inherent randomness. His hypothesis was that the universe was not governed by chaos but by order and hidden structure—ideas that resonated with his belief in a deeper, interconnected reality. This search for clarity and cohesion was more than just a scientific pursuit for Bohm; it was an expression of his confidence that the universe had underlying principles waiting to be uncovered.

Bohm's deterministic view of quantum mechanics came from his discomfort with the dominant Copenhagen interpretation. He questioned the idea that quantum particles only have definite properties when observed, as it didn't align with his belief in an ordered universe. For Bohm, this view implied a disconnection between reality and human understanding, a gap that contradicted his belief in a more integrated, holistic cosmos. He was inspired when thinking that there must be a more comprehensive theory, one that could explain quantum behavior without resorting to randomness or the limits of human observation.

Bohm found a kindred spirit in Albert Einstein, who also questioned the current probabilistic understanding of quantum mechanics. Einstein famously said, "God does not play dice," reflecting his discomfort with the randomness proposed by the Copenhagen view. The two men shared a philosophical affinity in their belief that the universe must operate by more predictable, deterministic laws, even if those laws had yet to be discovered. Bohm's discussions with Einstein were pivotal in shaping his commitment to hidden variables and determinism, further reinforcing his belief that deeper principles could explain quantum mechanics.

Intellectual Battles and Shaping Hidden Variables

Bohm's theory wasn't simply a rebellion against the established physics community—it was the product of intense debates with some of the greatest minds of the time. His ideas were shaped by his interactions with figures like Niels Bohr and Werner Heisenberg, both champions of the Copenhagen interpretation. Although Bohm disagreed with their views, these debates sharpened his thinking, pushing him to refine his theory.

One of the most impactful exchanges occurred with Bohr, a larger-than-life figure in quantum mechanics. In a series of discussions, Bohm expressed his discomfort with the idea that reality was fundamentally probabilistic. Bohr, in turn, defended the Copenhagen view, arguing that quantum mechanics didn't need hidden variables to explain phenomena—it worked fine as a predictive tool based on probabilities. Despite Bohm's respect for Bohr, he couldn't accept a universe governed by chance. For Bohm, quantum mechanics needed to offer more than just predictive power; it needed to make philosophical sense.

Another key figure was John Bell, who played a critical role in confirming Bohm's hidden variables theory. Bell's own work, notably his famous Bell's Theorem, showed that certain predictions of quantum mechanics could not be explained by any local hidden variable theory. While Bell's results complicated Bohm's quest, they also brought hidden variables into the spotlight. Bell admired Bohm's courage to challenge the status quo and recognized the merit of Bohm's attempt to develop a more deterministic quantum theory. Bell's theorem indirectly paved the way for renewed interest in Bohmian Mechanics, especially as physicists began re-examining the foundations of quantum theory in later years.

Even though Bohm's ideas were met with skepticism, his exchanges with these giants of science pushed him to deepen his conviction that the quantum world could be understood in a way that aligned with order and structure. These debates didn't discourage Bohm—instead, they propelled him forward, giving him the confidence and clarity he needed to fully develop his theory.

Bohm's Broader Philosophical Influence

Bohm's deterministic view of quantum mechanics was urged on by his broader philosophical outlook. Influenced by Eastern philosophies of holism and interconnectedness, Bohm was drawn to the idea that all parts of the universe were deeply linked. And that the apparent randomness of quantum phenomena was simply a reflection of our limited understanding. His later work explored the idea that reality consists of an implicate (hidden, underlying structure) and an explicate order (visible, unfolding one), where everything is interconnected as part of a larger whole.

The concepts of implicate and explicate order are central to David Bohm's later theoretical work, where he tried to explain reality's profound nature beyond the standard frameworks of quantum mechanics. Here's a quick breakdown:

Implicate Order:

- A hidden, underlying framework where everything in the universe is connected and enfolded together. In Bohm's view, this level of reality is not immediately visible to us, but it is fundamental.
- Think of it like a deeper layer of reality where all things are interconnected, even if they appear separate on the surface. In this order, particles (and by extension, all matter) are part of a larger, unified whole.

Explicate Order:

- The visible, "unfolded" reality that we experience in our day-to-day lives. It's the world of distinct objects, particles, and events that seem to exist separately from one another.
- This is the reality we think of in classical physics, where things have defined positions and properties, and interactions follow predictable laws like cause and effect.

In Bohm's framework, the explicate order "unfolds" from the implicate order, meaning the hidden, connected nature of the universe gives rise to the seemingly separate and independent things we observe. Bohm saw these ideas as a way to explain quantum phenomena and broader aspects of life and the universe, offering a holistic vision where the visible world is just a tiny part of a much larger, entangled whole.

This holistic view was central to Bohm's belief in hidden variables. For him, the "hidden" part wasn't just about missing information in a mathematical sense but was a reflection of the universe's layers of reality. These layers, Bohm believed, would one day be uncovered, revealing a coherent and unified understanding of the cosmos. His theory of Bohmian Mechanics was just one part of this larger vision, a step toward making sense of a universe that, to him, was ultimately knowable and structured.

The Philosopher of Dialogue

While Bohm's contributions to quantum mechanics are well known, his intellectual interests extended far beyond physics. His emphasis on philosophy made him stand apart from many of his peers. Bohm was fascinated by the nature of thought and communication, and he developed a theory of dialogue that aimed to foster open, non-judgmental conversation. His belief in dialogue was grounded in his understanding of how ideas evolve and change over time, and how people can break through fixed patterns of thinking by engaging in deep, collective exploration.

Bohm's concept of how to communicate had a philosophical approach to life. He believed that in order to address the world's most pressing problems—whether in science, society, or personal life—people needed to engage in meaningful discourse. This process involved listening without preconceived judgments, questioning assumptions, and exploring the underlying structures of thought. For Bohm, dialogue was a way to uncover truths, much like how quantum mechanics sought to reveal the underlying design of reality.

His time with Indian philosopher Jiddu Krishnamurti played a crucial role in shaping his philosophical ideas. Bohm and Krishnamurti shared a deep cognitive bond, and together they explored questions of consciousness, thought, and reality. Their conversations, spanning several decades, explored the limits of human knowledge and the potential of transcending conditioned thinking. Their relationship was a great example of the fusion of science and philosophy. Through these intimate talks, Bohm came to believe that the same cohesion he saw in the quantum world also applied to human thought and consciousness. For Bohm, understanding how we think and communicate was as important as understanding the physical universe.

Struggles with Isolation and Exile

Despite his brilliance, Bohm faced big personal and professional challenges. The political climate of the McCarthy era cast a shadow over his career. As anti-communist sentiment swept across the United States, Bohm became a target of the House Un-American Activities Committee (HUAC) due to his earlier involvement with communist groups. In 1950, he was arrested for refusing to testify against his colleagues, although he was later acquitted. Nonetheless, the damage to his career was severe. Bohm was suspended from Princeton University, and his association with communism made it nearly impossible for him to secure another academic position in the United States.

This forced Bohm into exile. He moved first to Brazil, where he took up a teaching position at the University of São Paulo. While this period of exile separated him from the leading scientific communities, it also allowed Bohm to focus intensely on his work, both in physics and philosophy. Bohm later moved to Israel and eventually settled in the United Kingdom, where he found a home at Birkbeck College in London. There, he continued developing his ideas on quantum mechanics and engaged in meaningful dialogues with scientists and philosophers from around the world.

The psychological toll of political persecution and professional isolation weighed heavily on Bohm. He struggled with feelings of loneliness and alienation, particularly during his time in Brazil, where he found himself cut off from the collaborative scientific environment that had nurtured his early career. Despite these hardships, Bohm's resilience was displayed. He continued to publish groundbreaking papers, including his work on the Implicate Order.

A Vision for Science and Society

Bohm's mental curiosity was not limited to the realms of quantum mechanics and consciousness. He believed that the entanglement he saw in the quantum world also applied to human society. Bohm envisioned a world where people engaged in meaningful exchange to solve conflicts and build a better society. He believed that many of the world's problems—whether they be political, social, or environmental—arose from fragmented thinking and the inability to see the natural law of reality.

This vision extended to his work with younger scientists. Bohm was an inspiring teacher and mentor, particularly during his time at Birkbeck College. His unconventional ideas attracted students who were eager to explore the foundations of quantum mechanics and engage in lively discourse. Bohm's holistic approach to teaching emphasized the importance of thinking critically and questioning established paradigms. He encouraged his students to look beyond the technical aspects of science and consider the broader implications of their work for society and human understanding.

Bohm's influence on the next generation of scientists and thinkers cannot be overstated. His ideas on the nature of reality continue to resonate in fields as diverse as psychology, social science, and education. Bohm's belief in the transformational power of conversation has inspired countless people to approach problem-solving more openly and collaboratively.

Bohm's Interests

Bohm found solace and inspiration in music and art. Classical music played a significant role in his life. Bohm often spoke of the parallels between music and science, noting how the structure and harmony of musical compositions reflected the patterns he studied in quantum mechanics. For Bohm, music was both a form of entertainment and a way of understanding the deeper rhythms of the universe. It helped him relax and think more clearly, providing a creative outlet that complemented his dynamic scientific work.

Art, too, intrigued Bohm. He was fascinated by how artistic expression could reveal truths about reality that science alone could not capture. This interest in the arts reflected Bohm's broader belief in the connection of all things—whether in nature, thought, or human creativity. His playfulness with music and art gave him a sense of balance and harmony, helping him navigate the intense demands of his scholarly pursuits.

Final Thoughts

For young scientists, Bohm's story is a powerful reminder that science is more than solving problems—it's about daring to ask big questions about the universe and our place in it. Bohm's perseverance, even in the face of isolation, his openness to new ideas, and his belief in the power of discourse highlight the importance of curiosity, resilience, and collaboration. As Bohm once said, 'The ability to perceive or think differently is more important than the knowledge gained.'

His legacy inspires us to see the oneness of everything—whether in science, philosophy, or life—and to remain committed to the pursuit of deeper truths, no matter the obstacles.

PART 1

CONCLUSION

As we conclude our look into the lives of iconic physicists—Einstein, Feynman, Heisenberg, Bose, and Bohr—the answers to the questions from the introduction become clearer. What drove these minds to push the boundaries of science? How much did their struggles, relationships, and values shape their discoveries or theories? Their challenges weren't mere obstacles; they were pivotal moments that fueled their drive for discovery, proving that scientific innovation is as much about the person as it is about the equations.

By understanding the human contexts in which these theories were born, we gain a richer and a lasting impression of the science itself. These are not just abstract principles of quantum mechanics, but the products of lives lived with passion, curiosity, and determination. When we see the humanity behind the science, the discoveries take on new depth, allowing us to truly understand their significance and appreciate the scientists' work.

Their lives remind us that learning science is not just about memorizing formulas or concepts. It's about seeing the world through the eyes of those who asked the big questions, who faced doubt and failure, and who remained determined to explore the unknown.

These giants have shown us the way, and now it's up to us to continue living with purpose. In the end, the greatest discoveries—whether about the cosmos or ourselves—are made by those who dare to think critically, communicate openly, and show their work.

PART 2

If someone says that he can think or talk about quantum physics without becoming dizzy, that shows only that he has not understood anything whatever about it.

–Murray Gell-Mann

Introduction

Second Edition, Part 2

It's Saturday morning. You have the day off, so you were able to sleep in... now what will you be up to? Are you hoping to get some gaming in? Turn on your TV and watch the latest cartoon or anime show on Netflix?

Whatever you decide, as you sit down, you may catch yourself wondering about the work that went into your favorite first-person shooter game or anime show. How many keyframes of motion went into animating the characters before your eyes? As you move your character through the video game, you might wonder how many commands and checks are happening at any given moment.

It's easy for us to take all these things for granted today, but the world of television and video games heavily relies on deceiving our eyes. Whether it is in the rapid blur of keyframes, or the dimensional 'magic' of 3D, technology and science have shown us that our eyes are all too easily misled.

Now, look down at your couch or computer chair and over at your TV and computer. What if you knew that despite how solid they feel, they are simply illusions of thought and energy?

Hold on a minute! You might be thinking, "Wait a second! I can see my computer. I can touch my couch. I can smell the flowers outside my window. I can hear my TV playing. I can taste the chips I'm snacking on. What do you mean this is all an illusion?"

Well, consider what you are holding in your hands right now. A book. What is the book made of? Perhaps you speculate its composition is made up of various elements like carbon, and you would be right. You might even know that elements and molecules are composed of atoms. What are atoms made of? Quantum physics has the answer: subatomic particles. What are subatomic particles made of?

Energy.

Our world, as solid and real as it might feel, is, in fact, made of constant flashes of energy. You can't see it, but it is happening—just like you can't see the individual keyframes of your character's animation in the video game you are playing. In short, quantum physics is a world within a world, where perception creates and impacts reality!

This sounds like something from a science fiction novel, doesn't it? Quantum physics, after all, is one of the newest and most innovative fields of modern science. Here, scientific pioneers are testing the boundaries of reality and challenging our ideas about cause and effect, time and space. The results of experiments and studies in quantum physics might surprise and challenge you. Still, in this book, you will have a head start on understanding and appreciating the amazing world of quantum physics. It's time for you to join us as we dive into the subatomic quantum world!

Meet Pantheon Space Academy

Welcome to Pantheon Space, a group of astronomy enthusiasts and scientists who love teaching and storytelling, especially about space and physics. Passionate about all things related to astronomy and science, we love studying space so much that our research doesn't feel like work. We have been engaged in the pursuit of knowledge of physics, astronomy, cosmology, and astrophysics since childhood.

Collectively, we study planets, constellations, and other space phenomena, but we also share a deep investment in education. Pantheon Space members are parents, teachers, and students who can share their research in an entertaining way that will increase retention.

At Pantheon Space, through knowledge and education, students build self-esteem, release the power of their imagination, and make significant academic improvements in science. Not only will our readers gain broad knowledge about the basics of physics, but they will also feel empowered by the possibilities of the future.

It's easy to think that studying is just a chore, but at Pantheon Space, we aim to provide material that a family will enjoy reading together. Learning about physics doesn't have to be just about getting the facts; it can also offer opportunities for quality time with friends and family and introduce an avenue to explore your imagination and relationships. In this book, Pantheon Space is ready to unlock the mysteries of quantum physics for the beginner!

An Overview

In this book, we will cover all of the fundamentals surrounding quantum physics. Starting with the basics in Chapter 1, we will look at a few simple ways of viewing and thinking about quantum physics as well as the history of this relatively new avenue for scientific study. In Chapter 2, we will tackle wave-particle duality to discuss the mystery of light before looking at the law of attraction in Chapter 3. Once we have a firm understanding of how to look at the universe as energy, waves, and particles instead of the solid matter that we are used to, we can begin to explore the theory of relativity, which is explained in Chapter 4. Chapters 5, 6, and 7 unravel the complex ideas behind Schrödinger's paradox and quantum entanglement, linking these discoveries to practical uses in the everyday world. What impact will quantum physics have on technology? Read Chapters 7 and 8 to discover the possibilities of faster computers crunching more complex numbers and potentially reversing the effects of time! Chapter 9 covers how quantum physics transforms our ideas about causality, and in Chapter 10, this book will showcase some of the most mind-boggling experiments to date!

Although physics is a scientific discipline that can take a lot of hard work, effort, and time to master, you can start the journey with this comprehensive—and fun—guide for a beginner quantum physicist. Let's zoom into the microscopic, ever-changing world of energy, light, and matter!

Chapter 1

The History of Quantum Physics

Nothing so much assists learning as writing down what we wish to remember.. –Marcus Tullius Cicero

Every morning, you wake up at some point. You might look out the window and notice the sun rising in the sky. Cars drive past your home. There may be flowers and trees rustling in the wind. Squirrels and birds dart in and out of the greenery. This is the living world that you are familiar with. Then, as the weekend approaches, you have a decision: should you go camping with your best friend or should you stay home?

You decide to stay home and clean. As you are sweeping your bedroom, you might be thinking that you are experiencing reality as we know it. However, what if, at the same time, there is an instance of reality where you went camping for the weekend instead? Can two realities exist at the same time? This is one of the original questions of quantum physics from when scientists began to study atoms in the early 1900s.

Looking closely at the world you live in, if you were to zoom in on the details, you would realize that everything is built from the same building blocks—atoms. However, depending on other factors, these atoms might be in different states, like gases, liquids, or solids. You are probably familiar with the idea of the three states of water: flowing water, ice, or steam. Whatever state it is in, water is still made of atoms, and when we look closer at atoms, we begin to realize that something strange is going on.

Atoms are formed of smaller particles, and not all of these particles behave as you might expect. Some of them change depending on whether we are looking at them or not. Some of them share the same space. What is going on here? Welcome to the world of quantum physics: the study of the very fabric of space, time, and reality!

Before we start to explore some of the most basic ideas found in quantum physics, let's have a look at the progress scientists have made over the decades since the idea of atoms, waves, and particles were first considered.

The Arrow of Quantum's History

Quantum physics has been developed over a relatively short period of time, but the roots of physics lay deep in the past. Perhaps it all began with the Greek, Indian, and Arab philosophers and scholars who developed math, the study of light, and ideas about motion. Historically, Greek, Indian, and Chinese philosophers proposed that all things were made of smaller substances, which one philosopher, Democritus, called an 'atom.' Over time, the works of these academics were translated into different languages, allowing more people to become interested and invested in science, like Galileo and Leonardo da Vinci.

Eventually, Sir Isaac Newton, who is often called the Father of Modern Science, began to formally theorize about gravity, time, and light. As a natural philosopher and mathematician, Newton first started to think about the invisible forces of nature. When the proverbial falling apple caught Sir Isaac Newton's eye, the famous scientist was inspired to research and analyze many scientific systems we take for granted today, such as gravity.

Back in the 1600s, technology could not provide as much information about the natural world as it can today, but Newton, resting on even older forms of knowledge, formalized and theorized scientific ideas that we call "classical mechanics" to this day. After Newton, chemists, like Dmitri Mendeleev, and botanists, like Robert Brown, made other important scientific discoveries, beginning the periodic table and producing evidence of atoms, respectively. Of course, although these theories worked well for hundreds of years, there were many inaccuracies that would slowly be debunked over time. Challenges to classical mechanics began with radiation and the discovery of atoms and electrons.

Setting the Stage

Today, we don't think a lot about radios since we often rely more on our phones. However, radio waves still play an essential part in scientific investigation and for a long time were crucial for long-distance communication before the Internet. James Clerk Maxwell played an important role in paving the way for the radio, but he also investigated electromagnetism, which inspired his theory of how electricity currents flowed. This led him to write about an 'ether' surrounding molecules in 1861, which wasn't really correct but began a serious discussion about what was affecting things like electricity.

Later in the 1800s, a very curious scientific couple, the Curies, began in-depth experimentation with radiation. Marie and Pierre Curie were invested in finding out how and why radiation happens. The very basics of elements were explored in their work. Thanks to their research, as well as the investigations of scientists like Maxwell, scholars developed more interest in the field of what would come to be known as physics. By this point, scientists knew that the world was made of particles, like atoms and waves, but these theories could be pushed even further!

Before you knew it, in the early 1900s, J.J. Thomson discovered the electron. Alongside Lord Kelvin, Thomson suggested that the atom was a ball made of various smaller elements jumbled together. Many call this the "plum pudding model," but you can imagine this concept about the atom like a fruitcake. Inside a fruitcake, you can find pieces of fruit mixed together still in chunks in the cake. This idea seemed practical at the time, but it quickly became obvious that this idea about atoms as a ball of random elements wasn't sufficient.

Other scientists, like Max Planck and Albert Einstein, got to work on the problem right away. To start with, Planck began to look at the problems posed by radiation. In the process of answering whether heat radiation was just particles, he created the formula known as "Planck's constant." At the same time, Planck began to suggest that the blackbody radiation he was analyzing wasn't just particles but also waves. As a result, Planck is often considered the father of quantum theory.

Albert Einstein, on the other hand, was looking at light. For a long time, light was only considered to be made of waves, but Einstein's experiments and observations proved that light waves appeared to be behaving like particles as well. As Einstein worked with high-frequency light, he began to use the word 'quanta' as a way to describe measurements in his experiments. This might be the first time the idea of 'quantum' was applied to physics. Still, Einstein and his fellow scientists had a long road of questions and challenges ahead!

Breaking the Boundaries of Science

To begin with, many scientists, including Einstein, struggled with the idea that radiation particles might also function like waves and light waves might also function like particles. From 1905 onward, Einstein continued to make breakthroughs in what we consider today as the field of physics. He hypothesized and observed the photoelectric effects, the impact of the observer on Brownian atomic motion, the establishment of reality atoms, and the theory of relativity. It sounds like a lot of work, but even at that time, the true understanding of the subatomic world of quantum physics was uncertain.

When I say 'atom,' you might imagine a ball made of smaller balls surrounded by other balls orbiting it. The idea of a nucleus orbited by electrons was designed by Ernest Rutherford in 1909. Since he shared his idea, we have used it as a way to describe atoms. However, there were a few problems with his diagrams, which Niels Bohr addressed in 1913. With his observation and explanation of atomic spectra, Bohr believed that electrons should lose energy, decay in orbit, and collapse into the nucleus of the atom. Electrons, Bohr theorized, couldn't just hold their position unless something else was going on. That "something else" coincided with Einstein's light quantum theory and Planck's constant research. Thanks to all three of these men, a better understanding of how electrons create an orbit while shifting positions to maintain energy, began to rock the scientific world. This became known as the "quantum of action," perhaps the first time we started seeing the use of the word 'quantum' in relation to physics.

Despite all of these advancements, the work of Planck, Bohr, and Einstein was just preparation for what was to come. Their work during this period was often called "the old quantum theory," while newer observations and discoveries were called "the new quantum theory." You might think that scientists would be excited about these findings, but both Bohr and Einstein struggled to accept the full ramifications of their discoveries. Let's find out why.

The 'New' Quantum Theory

Letting go of Newtonian classical mechanics wasn't easy. It was even more difficult to embrace some of the radical new ideas based on the work of Planck, Bohr, and Einstein. What if particles and waves were one and the same? What if two particles could be sharing the same space at the same time? What if just looking at the world changed it? This would challenge the very notion of science and the scientific method.

Erwin Schrödinger wasn't afraid to tackle the mystery. After taking into consideration earlier work by Louis de Broglie, Schrödinger proposed the famous Schrödinger Cat paradox. Thanks to his work,

Schrödinger was able to formulate the Schrödinger Equation for wave functions, which helped scientists understand the behavior of electrons. However, at the same time, Schrödinger also supported the bold idea that all matter is both wave and particle at the same time. Thus, the discussion formally surrounding wave-particle duality began with many physicists weighing in with their ideas.

At the Solvay Conference in 1927, 29 scientists, including Einstein, Bohr, Schrödinger, Heisenberg, and Marie Curie, gathered to talk about their findings. All of these very famous scientists were debating the foundations of what we know today as quantum theory. Not everyone agreed with each other. Bohr and Einstein were arguing quite a bit over Heisenberg's uncertainty principle and whether quantum theory can be used to truly understand physics. Bohr supposedly won, using Einstein's own theories against him, but the debate proved that quantum theory had a long way to go before being fully understood and explored.

One of the scientists at the Solvay Conference, Paul Dirac, was already making his own contributions to quantum mechanics. In 1926, he developed his own quantum theory of light and later in 1928, postulated theories on relativity. His theorization included a prediction of antimatter, which was later confirmed by Carl David Anderson in 1932.

From the 1930s onward, various scientists united to explore the boundaries of quantum theory. Many physicists worked on their own speculations on quantum mechanics. For example, David Bohm, based on the work of de Broglie, Schrödinger, and Dirac, developed another quantum theory framework with other scientists. In the late

1940s, Richard Feynman, Julian Schwinger, and Sin-Itiro Tomonaga formulated a complete theory of quantum electrodynamics to explain discrepancies in Dirac's theory. This sounds like a bunch of paperwork, doesn't it? It makes you wonder what the whole point of quantum mechanics is.

Quantum Mechanics Today

Since the initial breakthroughs during the early 1900s, the world of quantum mechanics has expanded to focus on ways to validate hypotheses, such as discovering that photons coexist in superconductors or confirming the existence of the Higgs boson particle at CERN's Large Hadron Collider in 2012. Still, this field of science has expanded beyond scientific inquiry and observation.

Although many physicists are just interested in figuring out how the world works, many have begun to apply their knowledge of quantum mechanics to improve technology. It would be a while before any of these theories would provide any practical uses, but various applications have emerged from studying quantum physics. For example, in the 1960s, Theodore Maiman was able to use the results of other scientists' research to build the world's first practical laser. In 2009, Aaron D. O'Connell invented the first quantum machine. More recently, in 2014, scientists announced that they had successfully teleported data between particles, which provides one more step toward the quantum Internet and data sharing. Later on in this book, we will take a closer look at other technological improvements that have been discovered thanks to the potential of the quantum world, such as faster computers and codes that are more difficult to crack.

At the same time, many people have suggested that quantum physicists are the world's modern philosophers. Not only does our growing understanding of quantum mechanics impact technology and other branches of physics, like theoretical physics, but our ideas about the relationship between cause and effect, the flow of time, or the solidity of matter are also being transformed. Quantum mechanics might one day affect how much we can transform our reality through thought!

CHAPTER 2

WHAT IS QUANTUM PHYSICS?

I think I can safely say that nobody really understands quantum mechanics. –Richard Feynman

It's a warm summer day, and the weather is looking good for our long weekend camping trip. We're starting our vacation off with a local game of baseball that you wanted to watch. Overhead, the sun beats down on the baseball pitch. The park stands are packed with parents and local baseball enthusiasts. Today marks the first day of the baseball season. The game so far has been close, but the home team is down by three.

This might sound unbelievable, but imagine you're the next player up to bat and you're walking over to take position in front of the catcher. With a player on every base, as you walk up to the plate, you know that you need to hit a solid home run for your team to be victorious.

The ball soars toward you. In the split second, as you swing, the universe splits into two. You don't see the split, but in that nanosecond, two words diverge—a world where you hit the ball successfully and a world where you strike out.

From there, the universe continues to split. We can label each possibility a different color to show how all of these probabilities branch off each other like a great tree. In the world where you won, which we can color green, you either go to an after-game party (blue), or return home to celebrate with your family (purple). As for the world where you lost, you might end up going home alone feeling depressed (yellow), or you go out with your coach to talk about new strategies (orange).

In the same way that twigs sprout from branches and branches emerge from the trunk of a tree, the possibilities generated by particles in the quantum world arguably spawn many possibilities and many worlds! From these findings, some scientists argue that quantum physics supports the idea of many worlds or a multiverse. What is quantum physics actually talking about?

The Definition

The potential reality of a multiverse might sound insane, but as we saw in the previous chapter, quantum physics revealed strange anomalies happening at the subatomic level. In the last chapter, you learned how quantum physics emerged as a scientific discipline, so you already have some idea what quantum physics is about. New discoveries surrounding the foundations of our cosmos changed over time, and this created the field of scientific investigation that we now know as physics. There are different kinds of physicists who look at various aspects of our world, such as astrophysicists, theoretical physicists, and quantum physicists.

Physics, in short, explains the mechanics of how our world works. It looks at invisible forces like gravity or thermodynamics; it is a way of observing and analyzing patterns to define and understand laws of nature; it provides the basis to form theories about how everything works together. Scientific work in the macroscopic or microscopic world can help to explain or predict natural cycles. However, quantum physics pushes at the boundaries of knowledge by going deeper than ever before.

Instead of using telescopes to look at the vast universe around us, quantum physicists use powerful technology to look inside the atoms of matter, which makes the world around us. These scientists are interested in looking closer at the interactions of particles on atomic and subatomic levels. As a result, quantum physics increases other scientists' understanding about the foundations of theoretical physics, chemistry, and biology.

Quantum mechanics provides scientists with mathematical frameworks that can help determine the behavior of particles on the quantum level. Bohr, Heisenberg, and Schrödinger among others led to the development of quantum mechanics. They were able to show how the placement or speed of a particle (or group of particles) can change over time. However, quantum physics requires even more work to be done.

Perhaps you have felt this way before when you are working on a puzzle. After pouring out all of the pieces, you should sort them by color and focus on completing the edges first. As you hold a piece of the puzzle in your hand, you know that it needs to work with other pieces to form a larger picture.

In the same way, quantum mechanics often combines with other scientific theories to create larger quantum field theories. You can imagine that this puzzle is made of four large pieces. So far, physicists have succeeded in fitting three of the four puzzle pieces together: electromagnetism, strong nuclear forces, and weak nuclear forces. Looking at the force of electromagnetism, we understand how atoms stick together and don't fall apart. Strong nuclear forces keep atoms stable, while weak nuclear forces reveal why atoms experience radioactive decay.

The fourth puzzle piece, the force of gravity, remains a mystery. Although many scientists understand how gravity works, there is no clear understanding of how gravity can be included in the working theories of quantum physics we currently have. This is the Holy Grail, the seemingly impossible task, before quantum physicists today. This theory, often called "the theory of everything," has been attempted by many physicists.

Today, we wonder if the theory of everything will ever be achievable. The late Stephen Hawking believed that this theory would never be fully explained or understood because we are limited by how we interact with and describe reality. Since quantum physics is difficult to explain or analyze, and the theory is not easily joined with other understandings of reality, many scientists have an uneasy relationship with the quantum world. Even Einstein struggled with accepting the lack of objective data in the study of quantum physics. In a letter to his fellow physicist, Max Born, Einstein is famously quoted as calling quantum entanglement a "spooky action at a distance" (Valdano, 2017).

Quantum physicists aren't giving up though. Today, they are still tackling some important questions about strange observations that are still not quite understood. Why do particles change into waves and back into particles again when observed? Why do wave particles occupy the same spacetime as other wave particles? Why do some particles affect each other over great distances? What role does gravity play in the behavior of particles? All of these questions are still up for debate.

The Five Basics of Quantum Physics

Although it might seem that there is ambiguity in quantum physics, there are a few things quantum physicists know for sure. These are important points to keep in mind as we go into more detail on the fundamentals of quantum physics.

First, The Cosmos Is Made of Wave Particles

Jumping into a pond in the summer is a totally different experience than jumping up and down on the icy pond in winter. In summer, when you cannonball into the water, you sink to the bottom before floating to the top. Around you, waves ripple out to the edge. In winter, the pond might be frozen almost solid, so all you will do is slip and slide around on its icy surface. The water of the pond has only changed its state from liquid to solid. To our senses, matter is interactable within the confines of three distinct states—solid, liquid, and gas. However, things start to look different in the quantum world.

One of the major takeaways to keep in mind is the realization that although we live in a world of solid matter, matter is not made of the pinpoint particles we have always imagined. Instead, particles, especially when not observed, transform into waves. As a result, many quantum physicists might say that the building blocks of the universe are a new category of matter that shares the properties of waves and particles. When quantum physicists found the Higgs boson, they often referred to it as a particle, but it is simultaneously also a 'field.' It could spread over an area of subatomic space, like a wave or a cloud. This can be a bit difficult to envision, but we will talk more about this in the next chapter.

Second: Measurements Are Always Precise and Distinct

When a magician pulls a rabbit out of their hat or makes a rose appear out of thin air, it might seem like magic, but most of us can eventually figure out how the illusion was created. During the magic show, we are entertained, but afterward, we can agree that magicians are skilled in the art of misdirection and staging, not in *actually* creating rabbits out of thin air. Quantum physics falls into this category as well.

Many people who talk about quantum physics might make it appear like magic or something that is just happening. However, the quantum mechanics that describe this subatomic world of transformation are describable with distinct mathematical frameworks that usually rely on integer multiples. If I say the word 'pi,' you might think of a dessert, but you also might think of the mathematical equation known as π. Pi, as in the number 3.14, is in fact much longer than two decimal places and goes on for millions of digits. In quantum mechanics, formulas do not multiply by unending numbers like pi. Instead, these numbers are very precise and measurable, which allows for atomic clocks to work efficiently.

As we explore the many different ways that quantum physics can surprise us it can be all too easy to forget that this field of science relies on precise measurement and complex calculations. Although quantum physics is often linked to randomness, probability, and uncertainty, most measurements used within this field of study require precise mathematical approaches.

Third: You Can Expect an Element of Randomness

What happens when you throw a ball up into the air? It falls back down to the ground again, doesn't it? Depending on your strength and the direction you throw it, the ball might take a longer or shorter time to hit the ground, but eventually, it will fall. Over time, you understand that the possibility that the ball follows the law of gravity is very high. Science is all about figuring out the probability of something happening. To do this, scientists create and test a hypothesis and attempt to define patterns of behavior in nature. This is intended to bring order and understanding so that we can better predict what is going on in the natural world around us. Objective data that can be repeatedly reproduced is preferred because continued experiments will reveal how probable something is. Quantum physics is rather different.

Thanks to various factors—like the observer effect discovered in quantum physics, where observers can change what happens in an experiment—quantum physics always has an element of random probability. Will the mass of matter turn into a wave or a particle? Sometimes, it isn't so clear. Using a symbol for wavefunction, which looks like Ψ (the Greek letter 'psi'), these formulas try to take into account the unknowns of quantum theory, like the lack of knowledge about a specific quantum object or 'real' but unquantifiable factors. Trying to figure out whether matter is in all states at once or in one unknown state continues to be a matter for investigation and debate. Until that mystery is solved, quantum physics has a strong element of uncertain probability.

Fourth: Spooky Action at a Distance Will Happen

Have you ever thought about calling someone only to have them call you a few minutes later? Or perhaps you have listened to your favorite music list, and you told yourself that you wanted to listen to a particular track, only for it to play right after. These kinds of connections might feel magical or spooky, but many would see them as simply coincidental. In the world of quantum physics, however, pairs of entangled particles, even when separated, appear to be connected over vast distances.

Quantum entanglement has been a source of interest and debate for many physicists, including Einstein, who tried to explain the cause by a "hidden variable." Although quantum entanglement is a proven observation, the hidden variable is not. Later experiments in the 1960s appeared to disprove the possibility of underlying causes for this phenomenon. You will learn more about this spooky side of physics in Chapter 6.

and Fifth, Quantum Phenomena Are Usually Microscopic

Let's go back to the first scenario of you getting up in the morning. Your toothbrush, shower, and breakfast are all very real interactions that you experience every day. The car you drive and the computer you use rely on a combination of science and technology, but it isn't obvious that quantum entanglement or wavefunction is happening at the same time. When you squeeze toothpaste onto your toothbrush or eat your food, you are interacting with the world, but your toothpaste will never intersect the same spacetime as your toothbrush. This is because most quantum phenomena are found on the microscopic level.

Why can't we see the flashes of wave-like behavior on a macroscopic level? Scientists are still trying to figure out that mystery. Wave-like behavior of matter that should be particles or vice versa is only found on atomic and subatomic levels. Looking for wave-like behavior in larger molecules has been an ongoing challenge. Physicists are looking at ways to use light or mirrors to observe the wave that creates the world we interact with through our five senses. Still, this is one question that hasn't been answered... yet.

Unraveling the Mysteries of Quantum

If we return to the baseball game scenario, you have to wonder what goes on in the other alternate realities. Do they continue forward on their own trajectories? Or do they last only for an instant? Since these alternatives are caused by every particle that makes up the cosmos, can energy sustain all of these countless alternate realities? What is really happening on the quantum level?

All of these questions are still being answered and debated since quantum physics is still solidifying its theories. Still, it's pretty exciting to think about what quantum physics has discovered and may continue to reveal in the future! Let's take a look at the basics in more detail before we dive into the deep end.

CHAPTER 3

THE MYSTERY OF WAVE-PARTICLE DUALITY

Not only is the Universe stranger than we think, it is stranger than we can think.–Werner Heisenberg

You find yourself standing on the edge of a small lake. Before you know it, you are picking up stones and throwing them, trying to make them skip across the water. On this clear, windless day, the only ripples are the ones caused by the stone. As each stone sinks into the lake's depths, ripples swell outward.

It's hard to imagine another part of reality where something holds the same qualities as the ripple on the lake as well as the stone in your hand, isn't it? Still, in the quantum world, that is exactly what is going on. On a fundamental level, the quantum world cannot be described simply as either/or, but both/and. Not only that but observation of space and time, when it comes to particles, can start to get a little weird.

One of these strange subatomic phenomena is called the wave-particle duality 'problem.' This wave-particle duality describes a condition discovered through quantum physics, where microscopic matter appears to have the characteristics of waves and particles at the same time. This discovery in the early 1900s transformed long-held ideas about how matter behaves almost overnight. It wasn't easy for many early quantum physicists to accept, but today, further experimentation, research, and analysis have proven the reality of wave-particle duality. We're going to take a closer look at the way matter actually behaves at atomic and subatomic levels.

Waves and Particles

What do you think of when you imagine a wave? Usually, if you draw a wave, it goes up and down like a rollercoaster. You can imagine that waves, like sound waves or ripples in the water, slowly dissipate and dissolve away. On the other hand, the stone in your hand feels solid. It might be round, flat, or a bumpy shape, but whether it is ground down to sand or as big as a mountain, the stone is still a singular object with defined edges and weight. As noted before, until the 1900s, most scientists believed that all of the matter that made the universe was formed by either light or particles.

To begin with, waves, like the ripples on the pond surface, might spread out in circles. Some waves, like light, can travel for long distances before dissipating. Sound waves, however, rely on tightly packed atoms in our atmosphere to travel, so they can die out more quickly unless they are particularly loud.

How about particles? Well, particles, like stones, have defined shapes. Remember Rutherford's model of an atom? It showed a bunch of small balls, electrons, zipping around a ball made up of smaller balls, the nucleus. The electrons, neutrons, and protons that make up the atom are particles. Since they are so small, they cannot be seen with normal microscopes. Although particles are microscopic, they are still objects that do not behave like waves—or so scientists thought.

The Mystery of Light

Isaac Newton and the following scientists between the 1600s and 1900s had many ideas about matter and light. Everyone believed that light was made of waves. However, Planck, Einstein, de Broglie, Compton, and Bohr all came to a different idea: light was *both* wave and particle! This turned all of science on its head.

Imagine a board with two holes drilled in it. These holes are right next to each other. A scientist called Thomas Young noticed that when you shine light through the holes onto a wall further away, you could see an interference pattern where the light waves overlapped. If you throw two rocks close together into the pond, you notice how some ripples overlap or cancel each other out. Young noticed that light did the same thing. So, light is just waves, right?

Geoffrey Taylor, in 1909, used the same experiment, but instead, this time, he took photographs of the light. After using a bright light, his special photosensitive camera showed that, just like Young said, there was an interference pattern. However, when Taylor turned the light down so that it was dim, the photograph showed that the dim light was made of particles. Eventually, the light filled up the area to make the interference pattern again, but it just took longer. Taylor's photographs showed that light didn't just make wave patterns, it made dot patterns as well! The light wasn't just a wave; it was also a particle!

Back at the pond, when you think about the ripple of water, it makes sense. After all, water is made of particles, which form water molecules. When the water molecules are disturbed by the stone, they will move outward and create ripples. At first, that might make sense for something like water, but what about the stone in your hand?

The Hidden Secrets of Particles

Look closely at the small stone in your hand; it's hard to imagine that it is also behaving like a wave. It's rock solid, and even if you ground it down to sand, the sand is still solid matter. Seeing wave behavior within particles on a macroscopic level, like a stone, is currently impossible, but things change when you look at the stone with powerful microscopes.

After Taylor's experiment, a few scientists began to wonder if the reverse was true about particles: Do particles also show wave behavior? De Broglie believed that particles might. After reading his supposition and further theories by Schrödinger, scientists tested it out. Using similar double-slit experiments, various scientists double checked particles, such as electrons. Their question was simple: Would electrons also behave the same way that the light wave particles did? The answer that came back from their experiments rocked the scientific world: yes! It turned out that electrons, when sent through the double split, showed themselves to be both wave and particle at the same time! From here began the debate as to how and why this wave-particle duality existed.

Quantum Physics Breakthrough

Quantum physicists like Einstein, de Broglie, and Schrödinger stepped in to further analyze and theorize why all matter was both wave and particle. Unsatisfied with just knowing that this was happening, quantum physicists pushed their understanding further, revealing more interesting secrets within the atomic and subatomic world.

For starters, Einstein began to look more closely at light. High frequency light, he discovered, easily showed that it was both wave and particle under certain conditions. Instead of thinking about light like a ripple on the water, Einstein saw light as being more like a shower of particles.

That's right! You can compare light to be more like rain falling from a cloud. Imagine a rain cloud as the lightbulb. When a cloud is full of water molecules, depending on how bad the storm is, the water falls in droplets. Depending on how powerful the storm is, how high the wind speed is, and how much water the clouds are holding, the water might feel like a light mist, or it might come down in heavy, bullet-like droplets. The wind pushing against the rain might make it look like it is falling in waves. In the same way, light sends out small 'raindrops' of light that increase with frequency and energy. Although the particles are considered separate, collectively they form waves.

By the 1920s, scientists had new approaches and theories about wave-particle duality. After all, since it was clear that both particles and waves held each other's characteristics, it was time to consider the idea that matter was in fact made of something else that had both wave and particle behavior. De Broglie and Schrödinger created new formulas to show what wave behavior a particle might have and what it might do in the future.

Building from there, a new quantum theory was formulated to describe the wave-particle duality. However, by that time, other ideas about causality and probability had entered the equation. This meant that not only was matter both wave and particle, but there was also an element of probability about how these wave particles behaved. Would they decay now or later? Would they act like a wave or particle in all circumstances? Imagine looking up at the sky and realizing that the cloud you thought would shower you with a fine mist suddenly bombard you with hard rain. Due to the fact that the new formula could only offer probabilities, certain scientists, like Einstein, who preferred classical mechanics or understandings of the natural world, found this quantum breakthrough difficult to accept.

Skipping Stones Across the Pond

The sky, trees, campfire, and smell of s'mores feels so real. It's so familiar, right? The sound of the stone splashing across the surface of the water and then sinking. The crackle of the fire. The rustle of wind through the trees. Looking around you at the small green-blue pond and the dark stones in your hand, it might seem surreal to think that the entire world is made of flashes of wave-like energy. Still, according to experimentation, scientific observation, and mathematical formulas, we now understand that the very basic building blocks of our universe hold two qualities simultaneously that form the tangible world around us.

CHAPTER 4

THE THEORY OF RELATIVITY

The supreme task of the physicist is to arrive at those universal elementary laws from which the cosmos can be built up by pure deduction. There is no logical path to these laws; only intuition, resting on sympathetic understanding of experience, can reach them.

–Albert Einstein

Roaring across the lake, the motorboat cruiser cuts through the water sending a wake of ripples and foam behind it. The wind whistles by, and it's hard to hear yourself over the growl of the engine. At 20 miles per hour, the boat isn't going to win the Grand Prix anytime soon, but the high speeds combined with the large splashing waves, makes this boat trip feel exciting. You might be hoping to get some water skis on and try out a few tricks.

Your friend at the helm waves at you. It looks like they may have thought of the same idea. As you make your way across the boat, you might be wondering what speed you are going. Making your way to your friend shouldn't take long, but you want to make sure you won't fall out of the boat. Are you going 20 miles per hour or less than a mile per hour? Are you going 67,000 miles per hour or 450,000 miles per hour?

What if I told you that all of those speeds could be applied to you? If you consider the context, it is true that you were moving less than a mile per hour as you walked over to your friend. However, to onlookers on the beach, you were moving at 20 miles per hour in the boat. To a space probe, you are moving with the Earth at 67,000 miles per hour around the Sun. If you didn't know, the Sun is also orbiting the center of the galaxy at 450,000 miles per hour! Recognizing the importance of context holds the key to understanding relativity.

What Is Relativity?

When you think about relativity, you might first start to think about math and physics, but let's have a look first at the word 'relative.' Relatives are a way for us to talk about family. As an adjective, if something is relative to something else, you can think of it as a way for us to compare qualities. Relativity, therefore, has deep roots in the idea of dependence or connections. In a similar way, relativity as a physics theory shows the different ways we can view ourselves, particularly in relation to time and space.

When it comes to measuring time and space, we think that we can measure time with clocks and space with rulers, but thanks to physicists like Einstein, we now understand that time and space are two ways of looking at the same thing. It's called spacetime. Before we jump into discussing Einstein's special relativity theory and what impact quantum mechanics has had on the theory of relativity, let's review what classical and general relativity is all about.

Classic Relativity

In the story above, you and your friend sitting in the motorboat as it zooms across the lake might feel stationary, but to others watching you from the beach you are moving fairly fast. You can judge your speed by the movement of the scenery around you. Even if the motorboat stops and you just sit there with your friend, the Earth, Sun, and even our galaxy are still moving.

As a result, classic relativity states that nobody is absolutely moving at one speed or absolutely not moving. Even when you are sleeping, you are moving, and if you are walking or driving your car, you are in fact moving faster than you think, thanks to how fast the astronomical bodies are moving. In short, our understanding of speed depends on what point of view we consider.

Special Relativity

Zooming around the lake in your motorboat, you might not realize it, but time is going slower for you compared to the swimmers on the beach. It's not something that you can feel because the difference between your motorboat and someone standing still isn't that huge. However, when we look at the speed of our Sun or other objects in space, the femtoseconds (a very small division of time) start to add up.

We will take a deeper look at this phenomenon later in the chapter. For now, you just need to know that although speed and other scientific factors might be relative to each other, spacetime and light remain absolute. The speed of light remains constant, whether you are standing still or moving very fast. However, as your speed increases, time slows down, which means that light takes longer to travel. This results in time dilation and length contraction, important concepts that formed Einstein's theory of special relativity.

General Relativity

Einstein wasn't going to stop there. He also wanted to explain what was going on when you sped up or slowed down your motorboat. In the process, he put forward a theory on how the force of gravity works. After all, Einstein's theory of special relativity worked well for objects moving at constant speeds in a single direction, but it didn't take into account speeding up, slowing down, or turning corners.

Taking these factors into account, Einstein theorized that spacetime curves around large masses that speed through space. The warping of spacetime could explain, for example, why the Earth is attracted to the Sun, or why the stars in the Milky Way galaxy orbit a black hole. Einstein's theory also explains why people feel weightless when they experience freefall. When you bungee jump, skydive, or float in orbit around the Earth, you do not experience acceleration through spacetime so you may feel weightless. Overall, Einstein's theory seemed to make a lot of sense, being verified by gravitational waves, black hole behavior, and the difference between time in orbit and time standing on Earth. Since then, some issues with Einstein's theory of general relativity have arisen thanks to quantum mechanics, which we will look at later.

Deeper Into Special Relativity

When we consider the speed of the motorboat and the speed of a person throwing a life jacket to another person on the motorboat, there are ways to calculate the actual speed of the life jacket being thrown. For starters, let's say the motorboat is traveling 20 miles per hour, and the life jacket moves at 2 miles per hour. To you and your friend, the life jacket is moving at a steady pace, however, things will look differently to observers on the beach.

If the motorboat is heading north and the life jacket is being thrown in a northward direction, Einstein's theory argues that to the swimmers at the beach, the jacket will appear to travel at 22 miles per hour. Check this equation out!

20 mph (boat speed) + 2 mph (life jacket speed) = 22 mph (total speed)

However, if the life jacket is thrown in the opposite direction, it will still be moving faster to the eyesight of the observers, but it won't be moving quite as fast.

20 mph (boat speed) - 2 mph (life jacket speed) = 18 mph (total speed)

This is because time, distance, and speed are dependent on who is observing what. However, when you add other universal constants like spacetime or light into the mix, things start to get a little more interesting.

Doppler Effects and Einstein's Equation

To begin with, let's say you have a siren attached to your boat. As you speed past the beach, you turn it on. Everyone is probably standing and shouting at you now, but they will hear something very different from what you hear. The sound of the siren as you get closer will get louder and higher in pitch (higher frequency). After you pass by and move on, the siren will become quieter and lower in pitch (lower frequency). This is known as the Doppler effect. It affects all wave types including ripples on the pond, sound, and light. If you had a light shining from your boat, the speed of the light wouldn't be altered, but how the light appears to us would vary due to its change in frequency.

From these observations, Einstein formulated his famous equation:

$$E = mc^2$$

Here, E stands for energy, m represents mass, and c is the constant speed of light. This means that the energy of an object equals its mass times the speed of light squared. Due to his discovery and resulting formula, Einstein and other scientists came to understand that energy and mass are different expressions of the same thing. Mass can become energy and vice versa. Therefore, if an object increases in energy, it also increases in mass.

Length Contraction

Another strange observation Einstein noticed was that the speed of an object affects how short or long they might appear. In our example, our motorboat is moving rather slowly, but if we were in a spaceship and able to move at the speed of light, length contraction would be observable.

Let's say that the length of our spaceship was around 50 feet long. Walking about the ship inside, it would still look and feel 50 feet long, but the length of the ship would look very different to someone floating in space. If they were able to slow down the video of your ship passing by them, the spaceship would appear to get squeezed as our speed increased. The observable length would maybe decrease to around 43 feet long, traveling at half the speed of light.

Time Dilation

With all of this relativity, it is important to note that the speed of light remains constant. Moving at 186,000 miles per second, light moves quite fast, but since it never speeds up or slows down, light became an important way for scientists to measure distance. This way, when astronomers discussed the distance between our Sun and other stars in the galaxy, they had an easier way to make calculations. Given how far and how fast the speed of light moves, it got Einstein thinking: What would happen if you flew at the speed of light?

Thanks to Einstein's special theory of relativity, we now understand that when you travel at very high speed, time slows down. In the case of our motorboat, it's impossible to feel the difference in time because the speed of the boat isn't actually that much bigger than a person standing still on the beach. However, if we achieve the speed of light, as mentioned before, you begin to notice the time loss in femtoseconds. Those easily add up over time. As a result, if we hopped into light speed and blasted away from Earth, time would slow down. Behind us, time would continue on in the usual way. If we returned to Earth, we might discover that our five-year journey was actually 100 years long for everyone else on Earth! Is this actually about time travel?

Sort of, yes! Theoretically, if you could travel faster than the speed of light, you could go into reverse time and travel back in time. Unfortunately, as noted with $E=mc^2$, the faster you go, the more mass you gain which requires a lot of energy to push. We have not yet figured out how to create enough energy to get someone to travel at light speed, so forget about going faster!

The Questions of Quantum Mechanics

For many physicists, Einstein's theory made a lot of sense. It explained so much! How do we calculate the speed of the life jacket being thrown on the motorboat? Why did the siren sound higher and louder when it got closer? What happens to time when the motorboat speeds up? How is energy and mass affected by the speed of light? Quantum physicists strive to unravel these questions starting at the subatomic level. The theories of relativity, particularly the general theory of relativity, made sense for motorboats, the Sun, or even black holes, but other strange anomalies found in quantum physics could not be answered by Einstein's formula and theories.

The Problem With General Relativity

Imagine a cloth stretched between a circle of chairs. If you set a ping-pong ball on it, the cloth doesn't change much, but if you put an apple or orange on the fabric, the weight of the fruit would make the cloth pull down. In the theory of general relativity, like the fabric of cloth, the gravitational field also creates a distortion around objects that are heavier in mass, like the Sun. Lighter objects, like our Earth, don't really impact the gravitational field a lot.

These gravitational fields spread all around us, but you could imagine that the field is closer to a grid with lines. These lines continue onward throughout the entire universe and guide objects through space. As a result, there are ways for scientists to use 4D geometry to calculate the direction of an object through the gravitational field or its effects on the field. When taking these fields into consideration, physicists and astronomers prefer to use 4D measurements. After all, as mentioned before, Einstein and other physicists agreed that space and time are two expressions of the same thing—spacetime.

However, the discoveries of quantum physics couldn't allow for a similar understanding of gravity. This is because quantum fields, which can be used to describe the other three fundamental laws of nature, do not follow lines. They are not as easily defined, nor do they always follow straightforward rules. We learned in Chapter 2 that quantum theory allows for an element of probability and not all particles follow specific behavior. On top of that, the quantum world is not easily described by 4D graphs or straightforward measurements.

Due to these fundamental differences, quantum mechanics cannot yet create a new theory for relativity or gravity. Quantum gravity remains a mystery. Some quantum physicists seek the "theory of everything" today. Yet, as noted earlier, many believe that an elegant formula that explains what Einstein discovered in space and what physicists discovered on a subatomic level may never be adequately calculated. Until then, our motorboat and its Doppler effect siren will continue to challenge quantum physicists for a long time to come.

CHAPTER 5

THE PARADOX OF SCHRÖDINGER'S CAT

We cannot, however, manage to make do with such old, familiar, and seemingly indispensable terms as "real" or "only possible;" we are never in a position to say what really is or what really happens, but we can only say what will be observed in any concrete individual case.

–Erwin Schrödinger

Is fishing boring? That depends on who you are talking to. I've got a few friends who enjoy pursuing the thrill of a big catch. The idea of sitting for hours watching your fishing line and bobber might appeal to some who enjoy sitting back and relaxing as they take in nature around them.

Watch the hook and bait sink beneath the gentle ripples of the small lake. After a few minutes or a few hours, you might notice that your bobber is moving up and down. Has it caught a fish? Or has it caught on some debris in the lake? You won't know until you pull on the line and drag out what you have caught.

At that moment, you don't know whether there is a fish or debris on the line, but you might logically think that it should be one or the other. For physicists, however, the quantum realm is not so easily understood. Until the moment you pull whatever is on the line out of the water, quantum physicists would argue you have caught both a fish and debris. That might sound nuts, but let's take a closer look at the idea of superposition.

Superposition

Superposition, an idea that early quantum physicists noted, might at first sound rather strange. How can you not know what a thing is? In the case of the fishing example, you know that the line will hold a fish or something else. However, quantum mechanics suggests that the object on the end of your fishing line is both until you look at it.

Schrödinger, Einstein, Bohr, and, later on, Feynman tackled ways to explain this strange quantum behavior. To begin with, as new models for explaining the composition of an atom emerged, the behavior of electrons was looked at very closely. It became clear through experimentation and mathematical calculation that electrons were not just particles, but also showing wave-like characteristics. This wave phenomenon could be explained by the movement of the particles as well as its spin.

Spinning sounds familiar. You can spin quite a lot of things in our everyday life, such as the wheels on a bike, the beaters on a hand mixer, or a globe on its axis. However, the spin of a proton is actually quite different because it is quantized. It isn't easily measurable, and the formulas and process for measuring the spin of a proton are very specific.

Overall, the variance of a proton's 'spin' can only be measured as being either up or down, left or right. Once observed, the wave-particle state collapses, allowing scientists to measure a particle's motion. Up until that moment, the wave-particle matter is in a state of superposition, where it could be said to be spinning both ways at the same time. Other kinds of subatomic wave particles, like protons or neutrons, may have more than two states that they share at the same time.

The Double Slit Experiment

The most famous experiment that repeatedly supports superposition is the two slit experiment. Imagine a wall with two gaps in it from top to bottom. Is your aim really good? Take a tennis ball and throw it through one slit. You can see that it hits a wall on the back. You might notice that you could choose the left slit or the right slit to throw the ball through, but the ball can't go through both of them at once. If we throw two balls at the same time, they pass through and hit the wall in similar ways. What if we filled this room with water? Looking at the area once it is filled, we could drop a stone in the water. You might notice that the ripples move toward the slits and create more ripples beyond, even though we dropped the stone on our side of the wall.

As we can see, the balls are solid objects that don't spread out. The water can create ripples on both sides of the wall, but its waves are less solid than the ball. As we know now, subatomic particles are both like the ball and the waves on the water, so when we shoot the particles through the slits, they often pass through as a single wave, but once they hit the wall, they only form a single dot. If we put a robot or camera to observe the particle, the wave-particle passing through the slit will go through like the tennis ball, as a single particle. Otherwise, it passes through as a wave.

Oddly enough though, even when the particles passed through the slit as particles, they still created interference-like wave ripples on the wall, showing that they still held wave-like functions, even though the wave was no longer observable. In some senses, the particles weren't in one distinct place at all, but potentially in other places at the same time. Since the past and future are not so easily observed and predicted, scientists of the day were challenged about their notions on superposition. As a way to model how crazy this was, Schrödinger came up with a famous paradox to show the main idea behind this form of quantum reality.

Schrödinger's Cat Paradox

Everyone may have heard about Schrödinger's cat paradox, but its relation to quantum physics has become confused over time. Quite a few people see this as a supportive story of superposition, but the reality is that Schrödinger, like Einstein, was struggling to understand on a scientific level what phenomenon they were dealing with on a quantum level. He formulated this thought experiment to explore the truth about what they had observed and needed to explain as quantum physicists.

The Thought Experiment

Let's get out a box, Schrödinger suggested, and put a living, healthy cat in it. We will also put in something that happens randomly. This quantum event would have to involve an unpredictable element, like uranium. If we put an atom of uranium in the box, it may decay at any moment. When the uranium's atom decays, a device measuring the atom would cause a hammer to fall. The hammer would break a small glass vial that held cyanide, resulting in the death of the cat.

If we closed the box and determined the cat would only die if the quantum event occured, we are not sure if it is dead or alive. This is all dependent on whether the atom has decayed or not decayed. As a result, two binary realities exist at the same time: The cat is alive, or the cat is dead. Until the box is opened, and we look inside, we cannot know for certain.

Usually, just like in the fishing example, we can always make a guess. On top of that, when we pull the fishing line out of the water, we know for sure that what we are holding is a fish and was always a fish, or not, depending on what you pulled out of the lake. However, in the quantum world, until observed, no precise calculation can be made. Furthermore, after observing the particle, certain qualities of the particle are lost and are not measurable. Therefore, the cat, like the subatomic particles unobserved by quantum physicists, exists in a state of superposition. It could be said to be both dead and alive at the same time. This leaves scientists in an unhappy position where they are not able to measure or carry out replicable experiments.

The Truth of Schrödinger's Cat

What many people might not realize is that Schrödinger's thought experiment was never carried out as a live experiment. This idea was posed by Schrödinger to show how ridiculous and unacceptable certain approaches to the superposition 'problem' had become. There were two divisions that arose out of challenges to quantum physics: classical approaches to science and more experimental approaches.

Classical approaches to science require researchers to follow specific steps to uncover patterns and laws within the cosmos with the aim to predict future events. As a result, verification of data requires experiments to be repeated with the same results many times. However, that is not entirely possible within the field of quantum physics. As a result, new approaches were suggested, which Einstein and Schrödinger struggled to accept.

One of the new approaches suggested that the observer played a larger part in the formation of reality. Since the wave behavior of a particle collapses when a quantum physicist observes it, they argued that perhaps previous to observation, the wave particle was spinning in both directions at the same time. As a result, reality was indeterminate, undermining on a philosophical level, the scientist's ability to gather data or even propose theories.

At first, Einstein and Schrödinger appeared to have lost the debate. Even Schrödinger's cat paradox has been used to prove what he was arguing against! From this initial debate, different schools of thought have emerged, not only on the phenomenon of superposition but also on fundamental debates about reality and the basis for the scientific method.

Quantum Schools of Thought

Some of the first philosophers also considered the world and attempted to theorize about why natural events, like storms or drought, happened. However, for hundreds of years, science and philosophy were divorced from each other with science focusing on the material world alone and only looking at "hard data." Out of these approaches to science, technology was able to flourish as well as improved forms of healthcare and medicine. Quantum physics began to look like it was returning to its ancient, philosophical roots when superposition and quantum entanglement entered the debate. Four main schools of thought in the scientific community arose out of this academic discussion, which we can still find today.

To start with, the "shut up and calculate" school of physicists focuses on measuring and data analysis. Instead of asking 'why,' these scientists prefer to stick to theorizing the outcomes of specific experiments and collect data. Instead of speculating, they seek to find answers within the material world, relying on more traditional approaches to science.

On the far end of this spectrum, the "create reality with observation" school of thought, often referred to as the Copenhagen interpretation, pushes the envelope of science into more experimental places. They argue that superposition and the collapse of the wave function that reveals the particle's spin proves that reality does not exist unless we measure it. This has led to some interesting theories about our impact on the world around us.

In between, two other methods exist that attempt to explain what is going on. Closer to the classical approach to science, "riding the wave," known as the pilot-wave interpretation, suggests that the particles are determined by a pilot wave. Formulated by de Broglie and Bohm, the idea was an attempt to solve the superposition problem with more classical approaches. Still, this pilot wave has not been discovered or proved, but if it were found, it would help determine the particle's position and state in a more easily verified kind of way.

Closer to the more experimental quantum physicists, the "many worlds exist" doctrine argues that when a quantum measurement is made, our observation results in only one reality of many. We might observe the spin of an electron to be up, but at the same time, another world exists where the spin is down. Just like in the example of the baseball game, these parallel universes branch off into alternate realities. This way, to a certain degree, the superposition continues, but also resolves in a logical manner.

As you can see, however, all schools of thought still face various challenges and obstacles in order to prove their position. For those in the "shut up and calculate" group, a world of potential research may be closed off. However, alternative approaches like "riding the wave" have not yet been proven definitively. Similarly, "many worlds" cannot be absolutely demonstrated, nor does the cosmos seem to have the required amounts of energy to maintain alternative realities. On top of that, the "create reality with observation" approach has spawned non-scientific exploration of philosophy and mysticism which not all scientists appreciate.

Overall, the phenomenon of superposition remains a challenge for quantum physicists. For now, the fate of Schrödinger's cat is uncertain. On a macroscopic level, however, the uncertainty of what is hooked on your fishing line is easily resolved. Our fishing trip might be successful, and your bait may have caught us dinner tonight!

Are you enjoying this book so far? I'd love it for you to share your thoughts and post a quick review on Amazon!

Chapter 6

Untangling Quantum Entanglement

Those who are not shocked when they first come across quantum theory cannot possibly have understood it. –Niels Bohr

On a warm sunny day when the wind is cool, the beach is one of the best places to be. I think that quite a few people would agree with me. Watching everyone shouting and playing makes you get excited for an afternoon of fun. What do you enjoy doing at the beach? Relaxing on the sand? Going for a swim? Building sandcastles? Playing volleyball?

If you look closer at those volleyball players, you might be able to recognize a way to think about superposition and quantum entanglement. You can see the guy bopping the ball and the girl on the other side hitting the volleyball to return it. The girl and the guy are several feet apart, but if we looked at their wrists, we might see that they have affected each other thanks to sharing the ball. Sweat, sand, or even microscopic skin cells might be carried on the ball from one person to another. What if the volleyball didn't exist, the two players still stood several feet apart, and they still continued to pass material to each other? That would look weird, wouldn't it?

In the macroscopic, everyday world, things cannot affect each other over long distances without a carrier, whether it is the wind, a blinking light, a shout, or even a thrown object. You can make someone sick across the room with your flu bug just by touching the waiter's hand. However, transmission depends on an intermediary object, like the waiter's hand. In quantum mechanics, particles appear to behave differently.

As we learned in the last chapter, superposition describes the reality before observation, where a particle may be in two or more states at one time. This made it hard for scientists to observe and predict data, but when they started affecting some particles, other particles being observed further away started to change. It was as if they were the same particle although they were separated. This phenomenon came to be known as quantum entanglement.

What Is Quantum Entanglement?

Let's say I take a white volleyball and a yellow tennis ball. If I hide them behind my back, can you guess which ball is in what hand? At this point, you could say that both balls are in an uncertain state, which is what Schrödinger noted. You could say that both balls are in both hands, a phenomenon we call superposition. As we learned in the previous chapter, many scientists believe that this represents a challenge to physicists and will be explained in time.

Suppose you guessed correctly: the white volleyball was in my right hand. That means that the yellow tennis ball is in my left hand. However, in quantum physics, observation has shown scientists that particles are always shifting and only can be defined when observed. So, you could say that once you observe this white volleyball, you freeze the state of the yellow tennis ball as well. This is part of quantum entanglement.

Quantum entanglement describes a moment when two or more particles share a quantum state. When one particle is changed, the other particle(s) also change in a similar way. As noted before, quantum physicists cannot predict whether the particle that emerges from a wave collapse will be spinning up or down or which way they were spinning prior to the wave collapse. However, as they measure the one particle, other entangled particles that share the same quantum state will mirror the particle being observed. If the particle observed is measured for vertical spin, it may show that it is spinning up. The entangled particle at a distance, when measured in a similar way, will begin to spin down. Whether they are a foot, a thousand miles, or a light year apart from each other, these particles somehow affect each other with no clear indication of how they did so.

Today, scientists use a variety of methods to look at subatomic particles. When it comes to entangled photons, scientists can recreate photons to experiment on. For example, high energy photons can be split into two lower energy photons that appear connected. That might make sense to a certain degree, since they came from the same source, but altering or measuring one of the twin photons will cause the other one to mirror despite the distance apart. Another method for entangling photons requires using a maze of mirrors. Quantum physicists will pass two photons through these mirrors without knowing which path they took, creating a good environment for entanglement.

Quantum Physicists and Entanglement

A long time ago, many people in Europe didn't understand basic facts of nature like we do today. Flies and diseases seemed to show up magically. It wasn't until famous scientists like Louis Pasteur came along that people realized that cause and effect was happening on a microscopic level. Once the microscope could analyze fly eggs or bacteria, the mysterious veil around flies and diseases was drawn back. In a similar way, a mystery shrouds quantum physics when it comes to superposition and quantum entanglement.

Einstein and Schrödinger, for example, recognized that the particles were behaving in ways unseen in the macroscopic world, but they were only challenged to find new ways to explain what they had observed. Instead of giving up, they believed that the quantum theories so far proposed were simply not good enough to explain superposition and quantum entanglement. Other scientists, like John Bell, however, decided to see if Einstein's theory was right. The results have spawned scientific and philosophical revolutions.

Einstein and Entanglement

Writing to Max Born, Einstein wrote that the behavior of entangled particles could only be seen as "spooky action at a distance" (Einstein & Born, 2014). Based on his understanding of space and time as well as the speed of light, Einstein couldn't figure out how the particles instantly changed each other. There was no lapse of time to show the distances between them, so he was faced with an unexplainable phenomenon.

At first, Einstein attempted to answer the question by proposing a local hidden variable. He was hoping that he could discover and prove an even more fundamental particle or state of matter that had imprinted hidden information into the particles. Even if the particles moved apart, the one would always spin up, and the other one would always spin down.

Take these balls, for instance. We could imagine a factory making both balls and no matter where these balls went, one would be yellow, and one would be white. When you were guessing which ball was in which hand, when you picked correctly that the white ball was in my right hand, it meant that of course the yellow ball was in my left hand. This is because both balls were white and yellow to begin with. The hidden variable inside them was determined by the factory. In the same way, Einstein thought that perhaps there was something as yet undetectable by scientists that had affected both particles simultaneously before they separated. As a result, a particle that was detected spinning up would have always spun upward, and its twin would have always spun downward. Physicists, however, couldn't find the hidden variable.

For the rest of his life, Albert Einstein invested his time in trying to reconcile the laws of gravity and its forces with the other three forces of the cosmos. His pursuit of the theory of everything left work for other quantum physicists to consider why particles were entangled and affecting each other's states.

John Bell and Locality

John Bell took up the challenge. There were some obvious issues with Einstein's ideas about hidden variables. For starters, observation had led quantum physicists, including Einstein, to agree that superposition meant that particles were in all states at once before observation. Also, no hidden variable had been discovered. Bell posed a simple, yet very complex mathematical equation known as the Bell inequality theorem to test whether Einstein's hidden variable could be at work.

Using statistical probability as well as mathematical formulas, Bell showed that particles didn't have predetermined identities and were not set to mirror each other 100% of the time. In our ball example, the tennis ball and volleyball appear to be separate and distinct from each other all the time. If they were entangled quantum particles, though, they would mirror each other with one spinning up and one spinning down. If you measured one and observed it spinning up with a positive value, there was a 100% chance that the other ball would be spinning down with a negative value. However, this only happens if you are measuring both balls for spin up value. Things became more complicated when scientists measured one ball for spin up and the other ball for spin right.

Bell suggested in his inequality theorem that if the balls had predetermined properties, when we test them separately for different values, the second particle wouldn't have a 100% chance of being opposite to the first ball. So, if we tested the white volleyball for spin up and positive values, and we tested the tennis ball at 45 degrees for a different angle of spin, both should have independent values in terms of positive or negative spin. That simply wasn't the case. It turned out that only when scientists were measuring entangled particles for the same spin direction was there 100% certainty that the particles would mirror each other. After measuring the two particles for different angles of spin, quantum physicists realized that the particles had a 50-50 chance of matching in spin.

There are many reasons why this phenomenon was observed, which are linked to mathematics, sine waves, and angles, but ultimately, Bell's inequality test has been used in many experiments, proving that where Einstein's hidden variable should have predicted a breakthrough in our understanding of quantum mechanics, nothing happens. Instead of statistically showing the probability of two particles mirroring each other, tests have proven that deciding what you are going to measure and at what angle you are going to measure it, will affect the entangled particles positive or negative spin.

Spooky Action at a Distance

When we look at the ways water or our tennis balls behave in the macroscopic world, we might feel like the quantum world has little to no real impact on our daily lives. After all, our tennis balls still behave like solid objects and water can still make ripples. Neither one looks like they are made of quanta that are both particle and wave. Neither one appears to have particles in superposition, nor do tennis balls and water show signs of quantum entanglement. Even when observed, quantum entanglement is hard to believe. Still, many quantum and theoretical physicists believed that the mystery could one day be solved.

Although no specific answer has emerged, the practical application of these observations to technology has already begun a new era of quantum computing. At the same time, philosophical debate surrounding reality has begun to heat up as we begin to come to grips with the idea that our world is less measurable and predictable than we imagined. You have to wonder how far you can push the boundaries of science and consciousness.

CHAPTER 7

THE LAW OF ATTRACTION

No Self stands alone. –Erwin Schrödinger

Now that night has fallen, it's nice to have a warm campfire going. As we sit in a circle, hot dogs are passed around with marshmallows, chips, and soda. Overhead, the sky no longer holds a hint of the Sun's glow. Now, the pinpricks of the stars and the cool light of the Moon shines softly. Only the warm glow of the fire lights our faces. The fire snaps and pops as you turn the hot dog skewer slowly. In this moment, everything feels so solid and dependable... and real.

Look down at your hand. You know your hands very well, right? Every wrinkle, every scar, every joint seems so familiar, you barely notice your hands until you experience an injury. Look at it a bit longer. What do you see? The protective layer of skin and nails which your body formed over muscle tissue and bones. Diving deeper in, you can analyze what your bones and muscles are made of—cells.

Go further in. What are the skin cells made of? Molecules. These are the framework for your hand, but even molecules are in fact arranged in patterns. Each pattern is different depending on the material or element. Skin molecule arrangements are different from nails, and so on. These complex systems evolved over time from atoms. As the building blocks of the cosmos, atoms are the most fundamental matter we can find... or are they?

I Observe, Therefore I Am

Atoms, as we have since discovered, are also made of subatomic particles, such as electrons, protons, and neutrons. Compared to atoms, molecules, and cells, which are also microscopic in size, reality appears to hold different rules on the subatomic level. These subatomic particles, as we learned earlier, are neither wave nor particle, living in a quantized state until observed. This has led some scientists to the startling theory that perhaps the wave-like state of subatomic particles forms a giant universal wave that is interconnected. It would explain superposition and quantum entanglement. This wave is more like a cloud or ether, and this means that the universe is, in fact, only formed of energy!

Since the observer effect causes wave particles to collapse, some scientists argue, like the Copenhagen school of thought, that reality is in fact formed through observation alone. That is, somehow our interaction with the world is keeping it together. Perhaps then, our minds are more entangled with the universe than we think. This is where the boundaries of science and philosophy begin to blur.

To start with, it is important to understand the battle between Einstein and Bohr over how reality can be measured and predicted. Although both were great men of science, Einstein was more invested in holding to deterministic, classical approaches to scientific methodology. This is why Einstein believed that particles were either affected by a hidden variable or held unidentified properties that caused them to reflect changes in each other.

On the other hand, Bohr had to be honest about what the scientific community had been observing: particles appeared to lose wave function and reduce to a position only when they were being observed. The observer effect had been proven.

Back then, Einstein technically lost to the Copenhagen school of thought. He suspected that the new theories of the day were simply not good enough, but more research needed to be done. Although Einstein's hidden variable theory seems to become less popular as a possible explanation for quantum mechanics, other theories, such as the pilot-wave interpretation, tried to tackle the subject in a more 'scientific' way. Certainly, Einstein was correct in the long-term, for Bohr's theory has since been refined and changed over time. However, the basic idea that our observations may have a deeper impact on reality resonated with scientists, philosophers, and mystics.

Entangled Philosophy and Science

On this beautiful night, we are hanging out by the campfire listening to campfire songs while roasting hot dogs and s'mores. Those chips look pretty tempting, don't they? I'll get us some. You like these ones, right?

As we work through the chips, it makes me wonder about how the universe works. It might seem strange to think about the mind and body during a time like this. Still, think about it. If we recognize that the world is built on subatomic particles that are simply expressions of energy, it gets you thinking about who we are. Our brains are no longer separate and distinct from the physical realm, but part of the same puzzle and linked to the same cosmos.

This is a stark difference to how many people viewed the relationships of the mind and body years ago. In ancient times up to Europe's medieval period, the world was believed to have two parts: the seen and the unseen. The seen world included these trees, the campfire, this hot dog, and this bowl of chips. The unseen realm included our mind and soul, and the land of the gods.

Then René Descartes came along: "I think, therefore I am." Sounds familiar, right? Summing up his new take on human consciousness and reality, Descartes' famous quote is still being talked about today. Now recognized as the Father of Modern Philosophy, Descartes tried to analyze the relationship between the mind (unseen) and the body (seen). Could something unseen and unmeasurable in the scientific sense have an impact on the physical, measurable world? Could two completely unrelated states impact each other?

Well, if we look at the food that our hosts brought, we can agree that the human body is quite different from the fruit, chips, meats, and drinks on the picnic table. Still, they are able to affect our body in both positive and negative ways. I mean, if I eat all of these chips by myself, I'm sure to put on pounds!

Believing that two separate states could in fact impact each other, Descartes argued that conscious will was the important factor. By our choices and ability to observe and analyze, we know that we are in fact alive and controlling our bodies. Although he agreed that the mind affected the body and vice versa, Descartes was able to formulate a fairly persuasive argument about reality and the mind. As a 1600s philosopher, Descartes didn't know about quantum physics.

After his death, a new great thinker stepped onto the scene—Sir Isaac Newton.

Thanks to Newton's work, other scientists and inventors were able to transform Europe and the world. Around this period and into the modern day, alternative ideas about how reality was formed entered the debate. Was it all just chemicals? Do people really have souls? Are our actions just the result of long millennia of evolutionary instinct?

All of these questions usually fell into two categories. Either the mind was just a product of the body's chemical processes, or the mind was linked to a more mystical, spiritual world. Quantum physics, particularly quantum entanglement and the law of attraction, entered the conversation with a bang. Although in some ways quantum physics made the debate a bit more complicated, its findings suggested that the previous ideas were affirming a false set of choices. Perhaps the mind had more agency in the universe than previously thought while at the same time being more interwoven into the fabric of the universe.

In the case of quantum entanglement, particles, although separated, appeared to influence each other. Einstein hoped to preserve objectivity and the ability to measure, which is necessary for most practical sciences, like medicine. However, the Copenhagen interpretation, formulated by Bohr and others, argued that the experiments in their labs proved that reality was formed by observing. Although the scientific reason for this phenomenon wasn't explained, it appeared as though the act of observation was forming the world.

Other scientists and philosophers, when they heard about this breakthrough, believed that perhaps our minds were able to form the world because they were sharing the same wave as the universe. Due to these discoveries in quantum physics, some philosophers and mystics have formulated alternative approaches to understanding and explaining reality as well as consciousness. This led to the concept of the law of attraction.

The Law of Attraction

According to the law of attraction, whatever we focus on will show up in our lives. If we focus on success and positivity, we are more likely to bring them into existence as we pursue our dreams and goals. If a person gets caught up in negativity and failure, those situations are bound to recur like a self-fulfilling prophecy.

Since quantum physics reveals that nothing is fixed, nothing is measurable, and reality is only defined by observation, then reality is just a state of potential. Living in a world that is simply flashes of energy that are formed through observation, it is no surprise that some people wish to harness this energy for themselves. If we apply our minds to that potential, we not only form the world around us, but we may be able to transform our lives and situations through our thoughts and passions. Instead of getting caught up in the machinery of the cosmos, we are able to create the world that we want to see.

We are part of the cosmos. We are energy and are surrounded by energy. We are also able to think and will ourselves not only into life, but also into a positive and happy one. In a world of fluidity, energy, and potential, how we view ourselves and our world can determine our future happiness, health, and success.

Does it sound too good to be true? On some level, like the Copenhagen interpretation, it feels like a crazy idea. If Einstein and Schrödinger struggled with accepting that the fundamental blocks of reality could only be subjectively observed, you can understand how hard it is for other people to also accept this possibility. After all, how many times have we wished for a better house or car? They don't just suddenly appear out of nowhere, right? In response, supporters of the law of attraction concept believe that our inability to determine the direction and aspects of our lives properly is a problem of education, focus, or misinformation.

To begin with, most people go through life making choices without really understanding why they are making choices or what is really at stake. With proper understanding and education about what is really going on around you in terms of quantum possibility, you will find yourself better positioned to achieve your dreams. For example, let's say you are choosing a job. Instead of just seeing it as the next step in your career, recognize that this moment is going to start you off on a branch of possibilities. Just like the baseball game example, there are multiple worlds with multiple possibilities, and you want to consider what branch you want to experience. Recognizing that this world is made of energy and potential is the first step to maximizing the power of the law of attraction.

Then, there is the question of focus. Many people might be going through life thinking that all they experience is out of their control. However, with the belief that you are able to affect your world, your focus may become sharper and help you achieve your dreams and goals more easily. For example, as you choose the job, when you weigh the choices, you have to consider more than, "What am I capable of?" Instead, you can consider what dreams you want to realize in the future, and then ask yourself whether that job works well for the journey you want to take through life. With the realization that your choices matter, you will feel more encouraged and empowered to make better choices for yourself.

Finally, misinformation can be a real problem, especially in terms of the false confirmation within yourself and from other people around you. A single negative experience or a piece of misleading information might cause us to create a false idea and attach it to something. Moving forward, you will be looking for confirmation and may end up reaffirming the negative expectations that you had begun with. Returning to the job decision example, you might have known an unhappy plumber. Perhaps your dad was unhappy working in the trades, or maybe you overheard someone say something negative about the trades. As a result, when you look at job opportunities, you begin to cut out potential avenues of employment out of fear or misinformation that may not actually apply to you in the present. Long-term phobias and anxieties are often created by repeated deleterious experiences caused by negative expectations and projections.

All of this sounds encouraging, right? It is easy to see why many people embraced this idea easily. That being said, like quantum physics, the law of attraction still has limitations. There are a variety of approaches to how to use the law of attraction in your life, but most agree that the art of manifestation requires time and energy combined with positive action and practical application. Let's take a closer look at manifestation.

The Art of Manifestation

Have you seen people chanting to themselves? Perhaps you have heard of people writing down numbers over and over again or seen people sitting and meditating. All of these are examples of manifestation. Manifestation is the practice of accessing the potential energies of the cosmos around you. It is a way to reorient your mind to get what you want by maximizing positivity and cultivating energy.

Whether a person believes in a specific faith or not, manifestation is often used as a technique to clear one's mind and focus on what you want. For some, numbers are considered sacred, so they either repeat to themselves or write specific numbers down in a journal. For others, words of affirmation or inspirational quotes can help them in manifestation meditations. In recent years, more resources for manifestation have increased, no doubt bolstered by people's openness to alternative ways of viewing the cosmos.

Not all practitioners view manifestation as a way to get what they want. Instead, for them, manifestation is simply a psychological way to cope and regain positivity. Instead of focusing on the negatives and traumas of the past, people might use manifestation and positivity-focused psychology techniques in order to gain energy for healthy change. Contrary to what people might expect, manifestation doesn't need to be based on unmeasurable factors, but founded on noted patterns in human psychology and behavior.

To be sure, manifestation and the law of attraction have attracted unwanted elements of predatory and unhealthy perspectives on reality. However, overall the law of attraction, as based on quantum mechanics, argues that the experienced realities of your past and present don't have to define the possibilities of your future. To many people, this theory or idea can be a source of positivity and empowerment, but there are some caveats to consider. Some realities about quantum physics, the macroscopic world, and the law of attraction just don't go well together.

Dreams and Goals Manifested

When Einstein was faced with the puzzle of quantum entanglement, he did not give up. Although his resulting hidden variable theory was less than satisfying, Einstein was invested in pursuing the truth. He might not be able to see his goal or even the full extent of the issue of reality, but Einstein continued to work on a "theory of everything" for the rest of his life. To a certain degree, you could say that even though Einstein never reached his destination, the journey he took provided invaluable information for the scientific community for years to come.

One of the key issues that plagued Einstein and still challenges quantum physicists today is the fact that the behavior of particles on a microscopic level cannot be seen on a macroscopic level. Although wave particles collapse into particles when observed, we do not see this collapse on a large scale. Furthermore, the properties described by Einstein's theory of relativity are divided when it comes to the quantum world. Although special relativity works well with subatomic wave particles, general relativity does not. However, general relativity is largely provable in the macroscopic world, and the subatomic world of quanta cannot be verified in objects larger than particles.

These two worlds, the microscopic and the macroscopic, appear to be working alongside each other, but neither of them can be explained as one whole. As a result, the pursuit of "the theory of everything" has continued to be a challenge. It also means that, scientifically speaking, there has been no definitive proof that microscopic laws work on macroscopic levels. While wave particles appear to be in both states before observation and, when entangled, affect each other instantly over great distances, these realities have no impact on the larger world.

Another fundamental issue lies in the misunderstanding of what actually happens during the entanglement process. Some philosophers and speakers discussing the law of attraction might compare quantum entanglement to magnets. From an outside perspective, magnets draw toward each other. It seems simple, and it works as a great example of the power of positivity. However, scientists understand that this is due to the force of electromagnetism that increases with the charge of negative and positive magnets attracting each other. In a similar way, particles also affect each other as though they are mirrored. As mentioned earlier, if a particle has been observed to be spinning up, when its entangled twin is measured the same way, it will always be observed as spinning down. This means that quantum mechanics isn't always about attraction of similar qualities, but that of defining probabilities and parameters for experiments, depending on the way a scientist measures the particles.

Still, the Copenhagen interpretation has opened the doors to begin revolutionary debates on how reality and the mind interact. If the entire universe is in a state of energy flux formed only by the observer, then it is crucial that we make choices to proactively choose the future we want to see. As we reorient our focus and energies, the many worlds of possibility can be more accurately channeled to bring us fulfillment and success. Quantum physics, however, does not ignore the real impact of the external world of waves that exist about you. The true lesson of the observer effect is that our external world impacts our focus and success just as much as our will does. As a result, we must, like Einstein and the quantum physicists of today, attempt to remain grounded in the world that we see as much as the world that we can't yet see.

Here, you can have this hot dog. While you think about what reality is, you can still enjoy your camping trip. These experiences are as meaningful as we make them, so let's focus on having a good time. Like the stars overhead, the mysteries of the quantum realm, entanglement, and the law of attraction remind us of possibilities and a world of research to explore.

CHAPTER 8

CAUSALITY IN QUANTUM PHYSICS

The weird thing about the arrow of time is that it's not to be found in the underlying laws of physics. It's not there. So it's a feature of the universe we see, but not a feature of the laws of the individual particles.
–Sean Carroll

A girl is running across the sand with a ball in her hands. Throwing the ball to a friend, she splashes into the water. Now, she's wet and everyone around her is wiping water out of their eyes from her big splash. Her friend now throws the ball further out, where it plops into the water. Left alone, the ball might slowly return back to shore, but someone else grabs it. A casual game of water polo has begun.

The game looks really fun. If we were sports commentators, our listeners would be enjoying the gameplay as well. Standing and watching the scene on the side, you take for granted one of the most familiar patterns we experience in life: the flow of time. Time, moving from past to future through the present, never seems to end. It also sets our understanding of how to order events in stone. If the girl hadn't splashed into the lake, her friends wouldn't have gotten sprayed in the face. If she hadn't thrown the ball to her friend, her friend would not have been able to catch it. In this way, we can say that cause and effect have a real impact on how we negotiate our world.

Without cause and effect, we cannot predict patterns in nature, like weather, nor can we deduce crimes or analyze history. Since we depend on the pattern of causality for order and security, including research and analysis, the reality that the quantum world had no apparent past or future did not sit well with many classical physicists, like Einstein. Superposition and the 'problem' of wave particle duality was bad enough, now other uncertainties arise due to entanglement as well. Could the commonly held ideas about time make sense anymore?

The Arrow of Time

Last night, we were skipping stones across the lake. Today, on the beach, we watched a swimmer making a splash. In both cases, the arc of the stones and the girl throwing herself into the water cannot be reversed. Their movement, like an arrow, was in a single direction through time and space. It's something that we are all aware of since all of us are getting older, but the more you think about it, the more weird it seems. After all, if we drive our car forward, we often can pull back in reverse. However, in other cases, like baking cakes, throwing stones, or diving into water, we cannot reverse our actions.

Arthur Eddington, a British astronomer, concluded in 1928 that time is asymmetrical. Moving in one direction, it does not appear to create balance between the past and the future. Rather, time is best shown through the second law of thermodynamics, the law of entropy.

Entropy explains how energy is wasted as it is transferred or transformed. For example, when you feel cold, you might want to make yourself a cup of hot chocolate. Over time, the heat (high energy) of the mug and drink seeps into the air and your hands. For a short time, it may make your hands warmer, but eventually, the heat will be swallowed by the larger amount of cold air. The only way to regain heat is to use another external heat source to warm up the molecules of your hot chocolate.

Like your drink losing heat, systems that are irreversible will tend to entropy, and in many cases, the process cannot be reversed. For example, you can't unscramble a scrambled egg. Due to the fact that the process of work expends energy, we can see that thermodynamic processes result in an increase in entropy. Things move from order to increased disorder.

On a macroscopic level, we see this every day. Over time, food goes bad. Soda goes flat. Clothes wear out. Cars get rusty. This tendency for things to age or get worse is an easy way of thinking about entropy and its link to time. Scientifically speaking, however, on a microscopic level, Eddington's work began a discussion on the creation and death of the universe. In the beginning of the Big Bang, scientists theorize that the universe was incredibly small and ordered with little to no waste of energy, but since then the universe has been expanding and the tendency to entropy is increasing.

Thanks to Eddington's questions, scientists and physicists began to also start thinking about entropy, time, and physics. This led many physicists to consider the issues surrounding causality found in the quantum world.

Quantum Causality

What state was this particle in before its wave function collapsed? Was the particle spinning up, down, or both before we measured it? When we measured *this* entangled electron, *that one* changed at the same time. Did the particle we measure cause the change, or did the entangled particle cause the change? How does quantum entanglement and superposition change our idea about causality and time?

According to theories and scientific proposals, quantum systems in superposition can be set up with a quantum switch. In this way, instead of being able to define whether particle A is affecting particle B, or vice versa, the relationship between the two particles is in superposition. You can think about it in terms of restaurants. You could say that more people are eating Chinese food because Chinese restaurants have been increasing in the neighborhood. At the same time, you can argue that Chinese restaurants have been increasing because the demand for Chinese food has caused people to go to Chinese restaurants more often. The idea is that sometimes it isn't so easy to tell which came first, the chicken or the egg.

What is retrocausality? The word stem 'retro' usually means 'behind,' 'backward,' or 'before.' Causality is linked to events that cause later events. Together, these two concepts form a ground-breaking physics concept: retrocausality. This describes the possibility that the future might be able to affect the past. This phenomenon is possible, as stated above, by using superposition and entanglement. After all, Einstein's "spooky action at a distance" might in fact involve entangled particles affecting each other due to an observer's decision.

This doesn't mean that anything actually travels back in time. Rather, signals might be able to travel to the past only between particles.

For some scientists, retrocausality might mean that time, at least on a quantum level, is symmetric and also reversible. Eddington's arrow of time wouldn't be discounted because time on a macroscopic level is tied to the law of thermodynamics and could be considered a special boundary condition. As a result, quantum retrocausality wouldn't affect the macroscopic world, just like other quantum states, such as superposition or wave particle duality are not seen in our day-to-day lives.

Other scientists believe that gravity might have a part to play in messing with time and causality. Just like we see around a black hole, time seems to slow down. This is called time dilation. It is possible that time dilation may be at work on a quantum level. This would cause time to speed up for one particle over another, creating the appearance of retrocausality. However, whether this is true has yet to be definitively proven. Recently, a thought experiment suggested that the gravity of giant planets or black holes might cause time to dilate on a macroscopic level, reversing cause and effect on some level. Communication between spacecraft could overlap due to time shifting with gravity. If scenarios like these are pursued, proven, and replicated in labs, various applications of quantum physics, like quantum computers, could improve in speed and efficiency. As a result, scientists consider this question to be one of the new boundaries of quantum physics research.

The Dream of Retrocausality

When you get sand in your ice cream cone or get too sunburned, you might wish you could travel back in time to make some changes. The ideas behind time travel can get very philosophical, which is why there have been many films and books about the dangers and opportunities to be found when time traveling. Do we want to travel back in time or forward into the mysterious future? It's a tempting question and certainly makes for some good stories!

So, can time travel happen? Not really. To go back in time, you would have to go faster than the speed of light. Just achieving the speed of light is impossible because with an increase in speed, your mass increases. Zipping through space at that speed, you would be squashed flat like a pancake! However, affecting the past on a quantum level has become one of the most exciting possibilities in quantum physics. As scientists continue to experiment on quantum particles and as our technological abilities increase, perhaps even the straight arrow of time might one day be reversed. Until then, we need to work on using quantum mechanics to advance our technological and scientific tools so that we are able to peer into the strange realm of quantum more easily.

CHAPTER 9

PRACTICAL USES FOR SPOOKY QUANTUM MECHANICS

Sometimes the public says, 'What's in it for Numero Uno? Am I going to get better television reception? Am I going to get better Internet reception?' Well, in some sense, yeah. [...] But let me let you in on a secret: We physicists are not driven to do this because of better color television. That's a spin-off. We do this because we want to understand our role and our place in the universe. –Michio Kaku

When our ancestors migrated in groups, they probably never have imagined a future where people would sleep in tents for the adventure of it. However, like us tonight, they gazed up at the night sky and wondered what was out there. Why did stars move across the sky over time? Why was the sun rising and the moon setting? Doesn't that group of stars look like they were making a shape? We might know the answer to these questions today, but the awe of the night sky remains. Nowadays, kids can recreate their favorite constellations with glow-in-the-dark stickers.

Did you know that glow-in-the-dark toys and stickers are the results of applying quantum mechanics to technology? The study of fluorescence and phosphorescence arose out of the quantum physicist's knowledge about the various states that elements could achieve. In the case of phosphorescence, if the electrons of certain elements are triggered, their spin can be changed, causing them to glow. Depending on the material, this change will take time to slowly go away.

Glow-in-the-dark objects, like the star stickers you put on your ceiling, are sensitive to the energy of light. Once the electrons are 'excited' (their spin states are changed), they will take some time to lose their phosphorescent light. All of this happens thanks to the exciting world of quantum physics!

Quantum Applications

Up until now, we have learned so much about quantum physics, but you might be thinking that this field of science is really up in the air. After all, quantum physicists can only theorize the state of a wave particle before they observe it, and when they observe it, further information is lost. Also, at this stage, physicists can only observe and measure what they find, but there are only a few things that are predictable, such as the response of an entangled particle. Still, beyond that, the precise state of the particle after observation is unknowable. Although the particle was measured and appears to be spinning up right now, it may return to its wave particle state after observation. Even the decay of certain particles is a randomized event!

Yet, quantum physics, despite its probabilities and unverifiable potentialities, has enhanced our lives already and will impact higher and more efficient levels of technology than ever before. Applying quantum mechanics to the field of biology, engineering, and the computer sciences, new advancements have been and will be discovered as we move into the future of quantum-based technology.

Biological Quantum Compasses

One of the things we have to be careful about when we go camping is getting lost. Out here, with poor cellphone reception, people can get turned around easily. Perhaps during our grandparents' time or before, people had methods to tell which direction they were walking in. Today, our reliance on technology has lowered our geographical awareness. For the rest of the animals, however, understanding direction is a combination of experience and instinct.

Can you hear them? Look over there. You might be able to see the familiar V-shape of a flock of geese flying south. Since it's the end of summer, the animal kingdom is already starting their yearly migrations. Many of us might understand why certain larger mammals may return to old hunting grounds. Their larger brain capacities lead us to believe that a combination of instinct and intelligence guide their migratory habits. However, birds are a different matter. How do they know in which direction to fly? Could it be that our avian friends know a thing or two about quantum entanglement?

Up until now, quantum entanglement is only able to be observed in labs. This is because you might choose a random wave particle to observe, but unless an entangled particle is also within the area of observation, you won't know if another particle is being affected by your observation. As a result, in order to observe quantum entanglement, I mentioned before that artificial entanglement has to be created by mirrors or splitting photons. It isn't easy to pull off.

However, the ability of birds to know the difference between directions might point to a biological compass that relies on quantum mechanics. Some scientists believe that birds have magnetic minerals in their beaks, which helps them become aware of the electromagnetic field which surrounds the earth. As a result, they can aim in the direction of a magnetic pole for migration. Another group of scientists believe, however, that a protein found in the eyes of birds might hold the key to seeing entangled photons.

In the eyes of birds, sensitive cryptochrome proteins receive information from photons entering the eye. As the bird looks around, the photons hit these proteins, exciting entangled electrons inside the bird's eyes. This allows them to see entangled radicals and forms a magnetic map for them to follow. If this is true, then it means that perhaps the quantum mechanics of the electromagnetic field around the earth holds the key to biological compasses.

Other animals, such as insects, lizards, or mammals, may also carry these cryptochrome proteins. Flies, for example, appear to have these same proteins, and their genetic makeup for cryptochrome proteins were also found in the human eye. However, before you get excited about seeing the magnetic field of the Earth, scientists have yet to understand whether it is or ever was a functional trait for humans. Perhaps humans needed help long ago to migrate, but today, we rely on other applications of quantum mechanics in order to get around.

Lasers and Holograms

Check out this laser pointer on my keychain. I often use this laser to play with my cat or draw attention to something specific on my lecture whiteboard. Lasers have actually been around for some time, so we feel really comfortable around them. Many science fiction nerds might be looking forward to a working lightsaber, thanks to *Star Wars*, but what many might not know is the important role that quantum mechanics played in the development of the laser.

Einstein first proposed some key ideas for the creation of lasers in 1917. He believed that the quantum theory of radiation might be applied to technology to create regulated light. This became known as "Light Amplification by Stimulated Emission of Radiation," 'laser' for short. Then, in 1960 the first working laser was prototyped. Theodore Maiman won the race to successfully create the first working laser.

Since then, lasers have been used for a variety of technological tools and innovations. Thanks to lasers, we are able to survey land easier. We can scan barcodes and cards, making payments and stock-keeping easier for day-to-day transactions. On top of that, computer technology like video game consoles, printers, or store scanners can store, read, and display information. Weak lasers today are used as cat toys and presentation pointers, but stronger lasers are used in scientific fields for medical work or astronomy. Technically speaking, all of these devices rely on the quantum world in order to work.

Today, one of the most amazing uses of light and lasers can be found underground, where fiber optic cables are laid. Using light to transfer data, applied quantum mechanics now supercharges the Internet. Housed in what we call fiber optic cables, light carries information much more quickly than usual cables do. Larger countries might find it difficult to provide this service for everyone, but some countries like South Korea enjoy super speedy internet from coast to coast thanks to this quantum-based technology.

The work of lasers isn't finished yet, however. In the future, it is possible that scientists will be able to stabilize intersecting lasers, leading to the creation of holograms. If holograms became common, our entertainment and computing world would be revolutionized yet again by quantum physics.

MRI Machines

What do you think would happen if the hydrogen particles in your body were stimulated? When data from quantum physics research emerged, many scientists from a variety of fields began to think of ways to use this knowledge for medical advancement. One such scientist, Raymond Damadian, came up with a device that used magnetic resonance imaging (MRI) to create clearer and more detailed pictures of our internal organs and tissues.

Up until that time, X-rays and other imaging techniques could give some idea of what was going on in our bodies, but MRIs revealed clearer images to aid doctors in medical prognosis and treatments. This is all thanks to a quantum process known as nuclear magnetic resonance. During an MRI scan, the magnetic force created within the machine affects fields of hydrogen, flipping the spin of hydrogen nuclei in your body. This creates distinctive patterns that form clearer pictures of your internal organs and tissues.

Considering what an MRI machine can do, now we know why they are so expensive! Not many people may end up experiencing an MRI, but this machine has become a crucial tool for revealing serious medical conditions like tumors, cancers, and other diseases in patients. Although it remains a mysterious science, quantum physics continues to support real-world advancements and provide quality of life improvements in the medical field.

Quantum Clocks

Do you own a watch? Most people don't these days because they rely on digital clocks on their phones. However, before the birth of computers and smartphones, most people relied on wind-up watches and classical forms of clocks, which depended on weights or batteries to run. Unfortunately, many of these clocks would lose the ability to tell time precisely and accurately due to wear and tear on their internal mechanisms. As a result, many mechanical clocks lose up to 10 seconds a day.

Digital watches are better, right? Not necessarily. Many digital watches rely on electricity that runs through a quartz crystal cut so that it looks and behaves like a tuning fork. The vibrations are counted by a circuit and can create a fairly stable account for time. However, temperature and pressure can affect many digital watches, resulting in a 30 second loss of time every month.

Today, our smartphone clocks are synced to receive ongoing data provided by a variety of clocks around the world. However, there is only one kind of clock that can provide the most reliable timekeeping ever—the atomic clock. That might sound quite dangerous, but atomic clocks rely on quantum mechanics to tell time extremely accurately. These clocks, being ultraprecise, will lose 1 second every 130 million years! What is going on inside this clock?

Atomic clocks, for now, are large scale clocks that are filled with atoms. The particles in atomic clocks hold radiation frequencies that can be measured as electrons switch positions in orbit around the nucleus of their atoms. As noted in previous chapters, electrons gain and lose charge as they change position. Since they change position and energy states at a regular rate, these clocks can rely on something that will never wear out, get dusty, or be affected by external environmental factors. That sounds perfect, right?

Still, scientists and engineers believe that even more precise atomic clocks will be created out of entangled atoms. With more atoms and the observation of entanglement, the data provided, and observation of particle changes will become even more precise, making current day atomic clocks look like digital watches!

Surprisingly, this isn't just about making the time on your smartphone more accurate. Our digital navigation, the Global Positioning System (GPS), relies on data to move at lightspeeds from you to different satellites and back. In order to keep the math, data, and timing accurate, satellites hold atomic clocks inside them. Now, your phone is able to keep track of where you are and where you want to go within seconds—all thanks to quantum mechanics!

Quantum-Enhanced Microscopes

As you can imagine, if atomic clocks that use entangled particles are going to be more accurate, how much more powerful could microscopes be if they relied on entangled photons? In Japan, advancements in microscopes rely on entangled particles. Once again quantum mechanics, practically applied, is helping to advance scientific progress.

Relying on a process known as differential interference contrast microscopy, these microscopes aim two beams of photon at something. The photons which are entangled respond instantly and reflect off the substance you are looking at. As they are reflected, they form patterns, revealing what they reflected off of. Since the entangled photons undergo changes instantaneously, they can reflect information even more clearly and quickly. Furthermore, the microscopic details of what you are looking at will be enhanced even more easily as well.

Overall, these quantum-enhanced microscopes will help medical and astronomical studies. Whether you are trying to uncover the secrets of the human body or of the universe, these microscopes will provide more detailed information and data. In the case of astronomy, for example, distant gravitational and light waves will be recognized more easily, allowing astronomers to observe and analyze large stars. If these microscopes provide more information with their entangled photons, even the gravity and light ripples of distant alien planets will be noticeable! Finding planets outside of our solar system is a very difficult task, but enhanced microscopes relying on entangled photons could open up the cosmos for a new level of exploration.

Teleporting Data

To get to this campsite, we had to drive two hours in the car. Since the site is so far away from home, we had to pack everything up carefully, checking and double-checking our lists. At some point in time, you might have wondered how much easier it could be if we could just teleport to the campsite park entrance. In the blink of an eye, teleportation would allow us to return home to get something we forgot. Unfortunately, teleportation is a long way off from being invented. However, scientists are already trying to tackle the early stages of teleporting data over long distances in the quantum world.

In a Scandinavian lab, scientists trapped electrons in diamonds and froze the diamonds so cold that the movement of electrons slowed down. This way, the electrons could be observed more easily. Then the scientists moved the electrons three feet apart. As they measured or observed a change in the first one, the scientists also noted an instantaneous change in the second one. It was good proof that quantum entanglement existed, but it was the first step to teleporting data as well.

Now that they are able to easily observe changes, the team will run further tests to figure out how far apart electrons can be and still change each other. Technically, the distances could be quite far. At the same time, the scientists want to figure out how and what information is being transported from one particle to another. If they do, we could figure out how to send information through entangled particles, making our communications inside and between computers superfast.

You know what superfast computers mean! Genetic sequencing, video game development, CG SFX rendering, and other computer related tasks wouldn't take as much time. In the next chapter, we will go into quantum computers in more detail.

Unbreakable Codes

What do codes have to do with quantum physics? Well, think about how important it is for us to have keys. We use keys to lock our homes, our cars, and our work offices. This way, our areas are kept safe from intruders. However, robbery and break-ins don't only happen in real life. In the digital world, your computer, your smartphone, and the applications they run have multiple levels of encryption. You can think of them as digital keys designed to keep your digital space free of hackers or other viruses.

Cryptography, the study and use of ciphers for encryption, has been around a long time. One of the earliest examples of encryption dates back to early 120 CE, when military commanders sent communications wrapped in a special way around a baton. This early form of cryptography is nothing like the encryption services we use today. When we send emails or texts to each other about buying more marshmallows or talking about the heat warning, it might not seem important enough to encrypt, right? However, big companies and corporations as well as international communications often require a lot of security. Security breaches could not only result in loss of important data but also might cost the company stock values, depending on the information leaked.

Practical applications of quantum mechanics may provide us with unbreakable quantum keys in the future. Quantum key distribution (QKD) relies on uniquely vibrating photons: A photon that is randomly affected so that its movement is restricted to only move up, down, left, or right. The message's encryption or sealed status couldn't be broken unless someone has a special filter to decipher the photon key and use an algorithm to interpret the message. Sounds tricky to crack, right? It is!

For starters, part of cracking a code is not getting caught or noticed. If someone tried to break the code, the photons would change, alerting the company that an attempt at 'reading' the photons had been made. Entangled photon particles are even better because anyone trying to hack in will be noticed right away due to the entangled photon responding instantly to change. Security breaches would be noticed right away.

Companies in Japan and Switzerland are already trying out versions of QKD today, but it will be a while yet before this form of encryption will be available for everyone. To begin with, the transmission of encrypted data is still limited for QKD. Since quantum-based encryption only works within 88 miles or so, this method of encryption, while very secure, is not practical for long distance communications yet.

Quantum Possibilities

After talking about the unknowns of quantum physics, you might have thought that the quantum realm only holds mysteries. While there are many unsolved and unanswered questions in theoretical and quantum physics, quantum mechanics has provided many scientists, engineers, and innovators with new possibilities. Already we take for granted the positive impact of discoveries made on a subatomic level. Quantum clocks, MRI machines, and lasers are part of our everyday life, supporting advancements in medicine and telecommunications. The field of research and development in relation to quantum mechanics has only just begun.

As you saw earlier, recognizing the power of quantum entanglement has led to cutting edge research on how to teleport data over short distances. Many of these experiments are still in the early stages of development, but these strides in technological advancements may lead to a future where computers achieve computation at mind-boggling speeds. Internet and rendering lag may become a thing of the past! What kind of world can we build with these kinds of computers? Let's find out!

CHAPTER 10

THE HOPE OF QUANTUM COMPUTING

A classical computation is like a solo voice—one line of pure tones succeeding each other. A quantum computation is like a symphony—many lines of tones interfering with one another. –Seth Lloyd

When you are out camping or hiking, you might feel really close to nature. You can start to imagine what life is like without a phone, the Internet, or even basic appliances. Simple tasks like gathering and preparing food become incredibly difficult. Surrounded by a vast wilderness of trees and small rivers, you might be able to catch a glimpse of wildlife you might not be able to observe at home easily.

Pulling out your phone, you take a quick snapshot of a young deer or a baby chipmunk. It's second nature to upload the photo to Instagram or Facebook, but halfway through the process, the upload times out. Your data network can't receive signals well, or perhaps you don't have any data at all. Suddenly, you might understand what life without a phone must feel like.

When we consider smartphones and computers, it is hard to imagine a time when we couldn't easily go out and buy one. Anyone older than 35 might remember a world before the first commercially available computers. During that time, telephones depended on lines, TV relied on cable, and your favorite games were Go Fish or Chutes and Ladders. Classical computing changed all of that within a decade, bringing not only innovations to mathematical and scientific research, but also enabling people to read digital books, easily write papers, and exchange ideas on the Internet. Later, understanding of data transfer allowed phones and video games to improve dramatically over time. As for the future, one thing is certain: Quantum mechanics might change everything.

Computers of the Past

01001000 01100101 01101100 01101100 01101111 00101100 00100000 01110111 01101111 01110010 01101100 01100100 00101110

Check out all of these zeros and ones! Do you know what these patterns are telling us? It might be hard to figure out, but to a computer, these numbers are actually holding information that we can access if we have the right tools. Understanding the importance of this code and how it relates to classical computers will help us recognize the amazing progress that quantum computers will be able to make. Let's start with reviewing the basics of how classical computers work.

Look at the code of zeros and ones again. When it comes to the computers we have grown up with and use today, all information is transformed into a code. This code is called the binary code. Each group of zeros and ones can hold different values. For example, 01001000 represents 'H,' while 01100101 means 'E.' Each number, letter, or data value we might use every day can be stored as binary code. That also includes pictures! In this case, the binary code above says, "Hello, world." Each of the pieces of binary used for each letter are called 'bits.'

Most computers today store and share data by using a complex assembly of metal and silicon. If you opened up your computer tower and looked inside, you can see a thin board that has various wires attached to it. This is a motherboard. Set into the motherboard, you can find a computer chip, usually kept cool under a special fan. Combining the computer chip with the motherboard and graphics card, computers are able to do complex calculations while storing, and showing data. All of this data is controlled by electricity and the behavior of certain metals.

Most electronics today rely on semiconductor-based technology that also relies on applied quantum mechanics. Research on a quantum level revealed how to manipulate the quantum behavior of solid objects, particularly silicon. Once scientists knew how to change the wave nature of electrons found in silicon, they were able to induce conductivity. By controlling these electrons, scientists and computer engineers were able to find a way to control electrical conductivity through circuits on a board.

Applying this knowledge, circuit boards were created that would respond to code. With a combination of software and hardware, the electrical circuits on the board would provide the basics of computing: input, storage, processing, and output. In this way, classical computers and phones help us to store and share data as well as calculate math or solve complex problems. Quantum computers, on the other hand, are revolutionizing computers from the ground up.

Computers of the Future

Let's say we want to make a tasty hot dog lunch. We have to get out the hot dog buns, the hot dogs, and the condiments. Maybe you like ketchup, mustard, relish, or onions on your hot dog. Perhaps you enjoy chili or cheese on top. The more complex you want to make your hot dog, the more space we need on the picnic table. Making a hot dog on a tiny plate with all those ingredients might be difficult, but with a larger table, you have more space to get your lunch ready.

The same is true about computers. In the old days, computers didn't have a lot of memory space, so it took them a long time to make simple calculations. Although we think that computers today are very fast, sometimes calculations related to math, astronomy, or biology take a long time. As a result, faster supercomputers are used to make incredibly complex calculations. If quantum computers can go even faster, it will take less time for scientists to calculate and analyze data. How is this possible?

Relying on foundational knowledge about quantum physics, these new computers aim to boost efficiency, storage, and speeds of computers. Instead of relying on binary code, quantum computers take into account superposition and quantum entanglement. By prompting two states on a quantum level, the computers totally revamp the binary system. These foundational pieces of data are often called quantum bits. However, unlike classical computer bits, qubits are both one and zero at the same time until they are measured. Using microwave signals, the particles are entangled and then flipped from zero to one. Microwave signals also control whether the particles are resting in superposition. From these processes, calculations or transformation of data can be made.

Even more exciting, if the machine has an increase in qubits, it also experiences exponential increases in calculation power. A two-qubit computer can make four calculations at the same time, but a three-qubit machine processes eight calculations! How many calculations could a four-qubit machine make? If you guessed 16 calculations, you are correct! This means that qubits can process information very quickly, but that's not all!

Inside the computer chip of a quantum computer, particles are kept super cold. Due to their frozen temperatures, they are more easily manipulated and entangled together. When these particles are entangled, the power and speed of the cubit is even greater, which means that these computers can crunch larger numbers than usual. At the same time, controlling the quantum states through the principles of interference, scientists can increase signals that are leading toward the right answer and cut out signals that lead to the wrong ones.

Each time a question or calculation is posed to a quantum computer, the answer isn't so simple. Quantum states, after all, are affected by probability. However, as the question is asked repeatedly, the answers change bit by bit and come closer to the theoretical or correct answer. Since quantum computers are so fast, reaching an answer doesn't take long. This means that research and data analysis could become easier for some scientists.

However, there are some drawbacks to quantum computers, at least for now. For starters, quantum computers today are still quite bulky. They are usually reserved for work in high-tech labs. Quantum computers have to be stored very carefully, protected from environmental factors like noise or dust. This is because maintaining certain quantum states is very difficult. Eventually, the information is lost as the effects of quantum fade through decoherence. Understanding how long coherence lasts is part of the process of using quantum computing.

Scientists hope that, over time, quantum computers will not only increase the speed of calculations but will be able to process more difficult simulations and calculations. Perhaps in a hundred years or so, everyone will be able to own and operate a quantum computer!

Work on Quantum Computers

In the past decade, different kinds of quantum computers have been made. A few invested groups have already begun work on increasing these computer's powers and building new coding frameworks. For example, in 2011, D-Wave Systems claimed to have built the first commercially available quantum computer. Two systems were called D-Wave One (128-qubit) and D-Wave Two (512-qubit). Controversy sparked over whether these systems were truly quantum since not enough entanglement was discovered among its qubits.

Other interested companies and groups, such as Google and NASA, have teamed up to research and develop quantum computing even further in a Quantum Artificial Intelligence Lab. In England, the University of Bristol made a traditional quantum chip available to the

Internet for free. It is connected to the cloud so that you can run simulations on and through it.

Due to work on quantum computers, many people believe that various fields of research will be boosted by the power of quantum mechanics. For example, simulations of weather patterns and climate change may be easier to track. As work on artificial intelligence (AI) continues, more complex systems of coding are required, which is very taxing for classical computers. Quantum computers, however, would be better equipped to handle the amount of data required to support an AI. Other fields of research, such as medicine, chemistry, machine learning, and engineering, also see quantum computing as an opportunity to push boundaries in fields of learning and discovery. Although many of these examples are still limited, computer science engineers and quantum physicists are excited to see how else quantum computers can be improved. The time for experimentation and discovery is far from over!

CHAPTER 11

STUNNING QUANTUM EXPERIMENTS

Experiment is the only means of knowledge at our disposal. Everything else is poetry, imagination. –Max Planck

Our camping trip turned out well. Not only was the weather great, but the lake wasn't too cold. We had a fun time swimming, fishing, and boating. When we got back to our campsite, we could light warm fires and enjoy a good time with our friends. Snacks and camp food might seem great, but over time, even when you are having fun, you find yourself excited to return home. Although returning to nature provided us with fun adventures and great memories, small daily tasks like showering, cooking, and caring for injuries got a bit more difficult. Ultimately, the convenience of our daily lives is hard to resist.

Microwaves, computers, and medical treatments, to name a few, are great examples of the ways that science has worked with technology to improve our lives. One specific field of science, quantum physics, has been particularly inspiring and empowering for innovators around the world. Thanks to quantum physicists, new discoveries, proposals, models, and ideas are explored every day. Some of these experiments support previous ideas and concepts. Others are starting a new phase of inquiry by proposing thought experiments or models. One thing is for certain: Quantum physics has only just begun!

Exploring Quantum Possibilities

The early days of quantum physics might seem like the most exciting. After all, there is nothing more shocking than disproving ideas that had been held as true for centuries. Einstein, Heisenberg, Schrödinger, Bohr, and their fellow physicists proposed many ideas, but it is only recently that many of their models have been proven correct. Even more importantly, the world of quantum physics is starting to yield amazing results within the world of technology and other practical applications as well. From revamping how the kilogram is calculated to bending time, ongoing experimentation and research in quantum physics continues to turn our world upside down. Let's take a look at what has been going on in more recent years!

Quantum Kilogram

How much do you weigh? In America, you might weigh 150 pounds, but the rest of the world would weigh you in at 68 kilograms. Here's the question though: How do you know how much a single kilogram weighs? Can our idea of weight change over time? If weight is affected by mass and gravity combined, how can we measure it?

In France, a cylinder weight of platinum-iridium was the final say on how much a kilogram weighs. However, what most people don't know is that over time even the atoms inside solid objects decay. Every now and then, an atom would be lost, suggesting that over time, this cylinder weight would no longer represent the kilogram as accurately as it should. On top of that, if you carried it up a mountain or placed the weight on the Moon, its weight would change because it is tied to mass and gravity.

As a way to counteract these inconsistencies, scientists have replaced this example of a kilogram with something that will last for billions of years. Based on quantum mechanical equations like Planck's Constant and Einstein's $E=mc^2$, the new quantum-based kilogram is mathematically explainable and relies on 10 to the power of 40 photons that are all wobbling on the same frequencies. Like atomic clocks, this new way of weighing the kilogram is more accurate. It also relies on our understanding of energy, making it understandable and recognizable wherever you are in the cosmos.

Entanglement Photograph

For us, cause and effect rules our lives. If we push this domino over, we can see how the rest of the domino line falls. Why the last domino in the line fell over isn't a mystery because we are able to observe the chain reaction. However, on a subatomic level, things are not so easily defined, as we discovered in our chapter on quantum entanglement. When two particles are entangled, as one is observed or measured, the other particle responds to the change as well. Since this seemed to break the objective world of measurement and no connection between the particles could be seen, Einstein and other physicists struggled to explain why quantum entanglement happened.

Today, we might not be able to explain quantum entanglement entirely, but at least we have photographic evidence that it does happen! At the University of Glasgow in Scotland, physicists unveiled the first photograph of quantum entanglement. After entangling two light photons, Paul-Antoine Moreau and his team split the photons and then caused the photons to go through four transitional phases.

Photographs were taken to show that when one particle was affected, the other one was also similarly affected. In these photographs, you can see that the photons aren't 100% the same in terms of angle, but their shape and rounded characteristics are the same! With experiments like these, scientists not only feel more confident about the reality of quantum entanglement, but also about the possibility of understanding "spooky action at a distance" even better in the future.

Tinkering With Retrocausality

Traveling back in time always sounds fun, doesn't it? Still, although the idea that our future might be affecting our past sounds exhilarating, time travel as a reality is a long way off. This has resulted in a variety of experiments, including a recent thought exercise that might push forward research into the mystery of quantum causality.

As noted in the previous chapter on quantum causality, superposition and quantum entanglement suggest that it is possible for the future to affect the past. However, no experiment has been successful at proving this effect. Recently, a new thought experiment in *Nature Communications* proposes that perhaps we might be able to see this phenomenon on a massive scale. Instead of looking for quantum causality at the microscopic level, perhaps the macroscopic world might hold a clue.

When it comes to supersize and gravity, you can't beat a black hole. In their thought experiment, Zych, Costa, Pikovski and their team argue that we might be able to catch quantum retrocausality if quantum superposition is combined with Einstein's theory of general relativity. Since objects with bigger mass can slow down time, time will move differently depending on your location relative to a massive planet or black hole. In this thought experiment, a supermassive planet might be orbited by two spaceships. It is possible that depending on the spaceship's position and experience of time, the planet will be closer and further from the spaceships at the same time. This could affect how the ships intercept missiles from each other, since who shot first might be affected by quantum gravity (Zych et al., 2019).

Since this is a thought experiment, this by no means proves that the future can in fact definitively affect the past, but it provides an important look at the role quantum gravity might play with superposition. Applied to quantum computing, this could help scientists figure out a way to use superposition during calculations. A system could analyze a chain of events as cause and effect at the same time, which would make calculations more efficient, fast, and powerful!

Metallic Hydrogen

In the night sky, Jupiter might not seem the brightest, but as the largest planet in our solar system, it stands out. Famous for its stormy winds and many moons, Jupiter remains a bit of a mystery. After all, thanks to its winds, poisonous gas clouds, and mass, getting close to the center of Jupiter isn't easy. On top of that, Jupiter is more electrically charged than usual. Although it has never been verified, scientists believe that the source of Jupiter's huge magnetosphere isn't just its super high rotation speeds. Deep in its center, hydrogen has been crushed by the mass of Jupiter until these particles become liquid metallic hydrogen.

On Earth, we are unable to find liquid metallic hydrogen even though hydrogen fuel is an efficient and easily stored source for energy. Many energy companies and institutions, including NASA, are invested in researching metallic hydrogen because of its powerful potential. If space travel is the new frontier, then creating rocket fuel out of metallic hydrogen could make space travel less difficult. Unfortunately, creating liquid metallic hydrogen has so far been considered impossible on Earth. This is because Earth doesn't have the same mass as Jupiter, so a special environment would have to be built in a lab to simulate or recreate high pressure.

In 2017, a team at the University of Massachusetts claimed that they had created metallic hydrogen, but their sample was lost during analysis. Recently, a rival team in France has claimed that they have proof of achieving metallic hydrogen. Paul Loubeyre and his team, working for the Commission for Atomic Energy of France, said they were able to recreate metallic hydrogen in their lab. The 2019 paper still has some issues, especially in terms of the remaining challenge of how to test this experimental material for electrical conductivity and its reaction heat. Still, Loubeyre and his fellow researchers created very innovative approaches to recreate ultrahigh-pressure environments, which will lead to more progress in the search for metallic hydrogen.

Entanglement of Memories

Only some people remember what life was like before the Internet was made available for commercial, everyday use. For some of us, memories of early dial up Internet remind us of how far tech advancements have come. Not so long ago, sending an email or a photo might take quite a lot of time and power. However, today faster and more efficient chips, motherboards, and memory cards (RAM) allow for increasing speeds, particularly in terms of using the Internet. On top of that, the emergence of "the cloud" has transformed how we use apps, store data, and access personal storage online. Quantum physics aims to take us even beyond that.

Although quantum computers are limited in size and abilities, work and research continues today with the aim to increase the power of computation, including Internet communication networks. We know that quantum computers are far from ready for public use, since the prime focus of most qubit chips programming is focused on crunching large numbers for scientific research and analysis. However, researchers are still experimenting on ways to begin the path to quantum internet.

In Europe and in China, two teams are looking at ways to use quantum entanglement and photon splitting to provide an alternative way to hold quantum information and create quantum memory. One experiment uses photons as the basis for their qubit computers. They store the photons and information in glass crystals that are filled with ions. Shooting the photons into the center of this crystal and ion model, it transfers energy and information, allowing the crystal to become a storage device. Now, they have a storage drive of crystal holding quantum information. Using quantum entanglement, they fire one entangled photon into the crystal and the other one down a fiber optic cable to another crystal in another room. Successful entanglements result in two sets of copied information in two separate crystals shared by entangled photons.

For many researchers, these experiments may be the start for a new approach to quantum computing and the Internet. This work may result in bringing us closer to quantum repeaters, which would be a fundamental building block for quantum internet. It might be a long while before we will see any form of quantum Internet, but the first step was taken in 2021!

Sixteen Futures

What is tomorrow going to look like? This kind of question can be answered in all kinds of ways. Whether you are thinking about the weather and what might happen at work or school, the future can only be described in terms of probability. As we progress through time, the possibilities narrow down to one set of experiences that we call the present. However, as we have learned, reality and time appear to be operating on a whole other level in the subatomic world of quantum physics. Recently, a quantum computer was able to calculate 16 possible futures at the same time.

In the baseball game story, there were a few different ways the game could turn out, and your choices afterward could then lead to other different life experiences. If life were about making a choice between two options, the futures would lead into 4 and then 8 and 16 futures, exponentially increasing in number. Understanding what those futures might become is an exciting idea.

In Singapore, Mile Gu and his team at Nanyang Technological University are using a quantum computer to simulate 16 different futures, using superposition. First, the team encoded a quantum computer with the possibilities created by flipping a coin 4 times in a row. Then, they shot a single photon down different paths at the same time. As the photon traveled, sensors took note of its conditions. After that, the team fired two protons and observed how their journey changed.

Using a classic model called "the perturbed coin," researchers tried to take into account previous realities that might impact future possibilities. Their model involved a box with a coin inside. Depending on how the coin was resting and whether it was shaken hard or softly, there were 16 proposed outcomes after 4 shakes. After synthesizing many factors, the team was able to use a combination of quantum computing and superposition to create a map of 16 possible futures.

Although it might not seem like a big deal, this experiment is a step forward to faster and more powerful quantum computers. Using protons in this way to predict possibilities could help AI use machine learning in order to increase accuracy of predictions. These computers would be able to create very accurate predictions for complex systems like the weather or the stock market. For now, we might not be able to travel to the future, but we can definitely start to shape what we want our future to be!

Mega Blast of Nuclear Fusion

Our ancestors burned wood for heat, but today we rely on a variety of sources for heat and energy. One of the major breakthroughs in the 1900s led to the implementation of nuclear power as a viable replacement for coal burning. With new developments and discoveries found in quantum physics, further research into alternative energy sources aims to improve the efficiency of hydro, wind, solar, and nuclear power. One step toward improved nuclear fusion was made in 2019!

In northern California, the Lawrence Livermore National Laboratory announced that they had broken a world record for nuclear fusion. The researchers, using 192 giant lasers, caused a small pea-sized portion of hydrogen atoms to ignite. More than 10 quadrillion watts exploded out with 1.3 megajoules of energy. That could power a trillion home furnaces!

The fusion reaction didn't last long—only 100 trillionths of a second, but it was an amazing step to improving nuclear fusion. Although the researchers are aiming to develop nuclear fusion weapons, other applications of efficient fusion are being considered. If energy from nuclear fusion can become more powerful with less radioactive material created, it would be able to fully replace traditional methods for generating electricity.

Going forward, other researchers and scientists are looking at the laser method as an alternative to create nuclear fusion. Many believe that this method may hold the key to create sustainable, green energy for the future, but the process used by the researchers at LLNL would have to be tweaked. Perhaps laser fusion energy programs are in our future!

The Quantum Turtle

Everyone likes a good fireworks show. Some of the best fireworks today can form the most interesting shapes, whether that's flowers, birds, or flags. For quantum physicists, though, the best quantum fireworks were observed at a subatomic level, and a mysterious pattern emerged!

In the last decade of research, quantum physicists realized that if you freeze an atom at super cool levels something strange will happen when you zap it with a magnetic field: quantum 'fireworks' happen. It looks like rounded jets moving away from the center, just like a firework. That seemed super cool, but at the University of Chicago, physicists believed that there was a pattern to discover.

Using machine learning, they recreated the experiment over and over, taking and analyzing data with the help of machine learning. By using a pattern recognition algorithm, the researchers were able to show that the atoms were forming a distinctive pattern after all—a turtle!

It's not really a turtle, but the shape of the pattern does look like a turtle if you were looking at it from the top down. There is a circle where the head might be, and another circle where the tail might be.

These circles are attached to a larger circle that could be the shell of a turtle. On either side of the circles, by the head and tail, four white dots formed, where the feet of a turtle might show.

Now known as the quantum turtle, this phenomenon has become yet another interesting phenomenon to explore in quantum physics. Why does this pattern happen? No one is sure yet, but scientists hope to use the algorithm method in this experiment to understand particles better. As the mysteries of quantum particles are solved, quantum-related technology, like quantum computers and networks, will increase in efficiency and power.

Bending Time

Whenever you drop your favorite mug or your new smartphone, seeing the shattered pieces or screen might make you wish that you could reverse time. The reality is that on a macroscopic level, reality dictates that many things cannot be simply reversed. You might be able to glue your mug back together or get your smartphone screen replaced, but it isn't 100% the same anymore. All of this points to the arrow of time, the laws of thermodynamics, and the unmovable reality of cause and effect. Of course, by now, you know that things are looking different on a quantum level.

A team of American and Russian researchers published a new study in *Scientific Reports* showing how they used a quantum computer to bend the arrow of time just a little. This quantum computer was very small, with only two qubits to perform calculations. Its job was to provide an environment for the experiment.

Based on common understanding of wave particle duality, a specific particle can be discovered at any point in its wave function. Over time, however, the wave ripples move outward, creating larger and less calculable possibilities of where the particle might be. It becomes more difficult to figure out where the particle might end up. In this experiment, the scientists were able to pull the expanding ripple of a wave particle within a very controlled quantum environment.

The computer had to be very small, so that it could control all parts of the system and environment. Keeping time reversal clear of entropy was difficult, so the computer was only successful 85% of the time. Still, scientists see this as a good first step to improved simulations in the future, as well as providing an alternative way to tackle the challenge posed by the arrow of time.

Cracked Quantum Tunnel

When it comes to the quantum world, simple things like throwing tennis balls at a solid cement wall has interesting effects. On the macroscopic level of reality, when we throw a tennis ball at a wall, it bounces back to us. However, when particles collide with barriers on a subatomic level, they appear to pass through. This strange ghost-like behavior might seem strange, but scientists kept working to figure out what was going on.

Many scientists believed the key to the answer lay in the characteristics of particles, like electrons, which were no longer considered simply particles but in fact a form of wave particles. This would cause interference patterns from the particles to form on the other side of the barrier. One way to think of this is to imagine a thin wooden wall standing in water. If a really heavy ball bounced against this light wall, the ball's impact would cause ripples to happen on either side of the wooden wall. However, when it comes to wave particle duality, scientists noticed something else in recent experimentations.

Various studies and research around the world have proven that particles in the right conditions can ghost through barriers! Thanks to wave particle duality, these particles may appear at any point in the wave when they 'land' on the barrier. If the barrier is thin, the wave extends beyond the barrier, which allows the particle to appear within the wave on the other side. This means that the particle can pass through the barrier since part of its wave passed through the barrier.

Once this phenomenon was caught in camera, many physicists wondered how long it took for a particle to break through the quantum tunnel. Recently, Aephraim Steinberg and his team at the University of Toronto have been using special clocks to time rubidium atoms racing through a special barrier they erected—a repulsive laser field. Their findings suggest that thicker barriers may encourage atoms to speed up as they tunnel from one side. Some scientists believe this might mean that atoms may be able to go faster than the speed of light. While work still continues on quantum tunneling, answers about the behavior of wave particles continue to challenge old assumptions about the abilities of wave particles.

Quantum Action in DNA

Every morning when we get up and look at ourselves in the mirror, it's hard to imagine that our body is changing on a molecular level all the time. As each day passes, our DNA copies and rebuilds itself. However, this process doesn't always work perfectly, and mutations can emerge in DNA. Although many scientists understand how genetic mutation happens, quantum mechanics might be able to explain why mutation occurs in DNA. After hearing about particles leaping across barriers or appearing in spaces where they shouldn't usually be able to go, it should come as no surprise that this might affect our DNA as well.

In 2021, researchers published a study in *Physical Chemistry Chemical Physics*, where they explored the possibility that quantum tunneling may be the cause of point mutation. The positively charged protons inside DNA might suddenly move over to a different portion of the DNA helix, causing errors later on when the DNA copies itself.

The early stages of research support this model, but the calculations and computational power of the team were quite limited compared to the complexity of DNA. Until computers increase in computational power, quantum physicists and researchers will not be able to truly understand how often proton tunneling might happen, how long a proton must be unstable in order to leap, and what conditions cause proton instability to begin with. Moving forward, further experimentation now focuses on increasing the DNA used in the experiments. In the process, the remaining questions about quantum instability in DNA might one day be answered.

Heat Making a Leap

Is space hot or cold? Many people might think that since space is largely 'empty,' it is cold. Certainly, as you travel further away from the Sun, planets become increasingly colder. However, heat can in fact travel through space, which is why the Earth maintains its life-giving warmth thanks to its orbit in a safe zone around the Sun. Radiation can cross the vacuum of space and provide heat to our planet. In this way, we both rely on the Sun for life and at the same time, must be careful to protect ourselves from harmful radiated particles.

On Earth, heat is shared through conduction and convection as well as radiation. When you hold a warm cup of hot chocolate, conduction is happening. Your hands are slowly warmed through the mug which is also warmed by the hot chocolate. This process of heating is often quite slow because the molecules of solid objects don't move as much as molecules in liquids or gases do. On the other hand, convection happens when energy is transferred through fluids. Often convection circulates, whether it is happening in the sky to create weather, or it's happening at home when we are heating or cooling our rooms. In both cases, cold air is denser, so it sinks to the floor or bodies of water. As the cool air moves across the floor or land, it becomes warm, thanks to the heater or Sun. As it gets warm, the air rises to fill the upper portion of the room. This cycle repeats slowly over time. For quicker heating processes, though, radiation is best. Radiation occurs when heat moves through electromagnetic waves, like the heat that comes from a campfire. Of course, all of these laws are upended on the quantum level.

At the University of California, Hao-Kun Li and fellow researchers demonstrated that heat could leap through a vacuum. Placing two super small drum-like membranes in a vacuum, the researchers heated the membranes up to different temperatures. The hotter membrane vibrated faster than the colder one. Together they caused quantum fluctuations to happen. These fluctuations pulled the heat toward each other. Eventually the vibrations began to share the same rate, as the hotter membrane moved heat to the colder one.

Since heat shouldn't be able to move through a vacuum without the help of particles, like photons, the research group believes they have finally proved the Dutch physicist Hendrik Casimir correct. The Casimir effect, a proposition posed by Casimir in 1948, suggested that quantum fluctuations in energy could pull objects toward each other in a vacuum. It looks like Casimir was right! Scientists believe that this experiment could help nanotechnology easily address heat related obstacles, creating new ways to keep circuit boards cool while running technology at super speeds.

Super Superposition

Can a person be in two places at once? Not yet, but if it were possible, it would make for some interesting situations! Maybe you'd be able to send a copy of yourself to school or work and just chill at home for the rest of the day. On a quantum level, subatomic particles, thanks to experimentation on the properties of superposition, seem to be able to do just that!

We know that thanks to wave particle duality, the state of a particle is both wave and particle at the same time before measurement. On top of that, any kind of positive or negative spin is unknown, so the particles are seen as having both spin states at the same time. When fired through the double-slit experiment, scientists noticed that particles created wave-like patterns on the back wall. Even when the particles were examined and lost their observable wave functions, they still formed wave patterns, which meant that the particles on some level were 'separate' yet still the same particle.

All of this was unsettling for physicists because although this phenomenon was identifiable on a subatomic level, this had not even been seen on a molecular level. Molecules, being more complex structures of atoms, have short wavelengths which are harder to detect. If a molecule is very complex, with around 2,000 atoms, their interference pattern is very tiny, smaller than the width of a single hydrogen atom.

Fein and his fellow researchers published a paper in 2019 stating that they had successfully proven quantum interference on a very large molecule. These molecules are called "oligo-tetraphenylporphyrins enriched with fluoroalkylsulfanyl chains" (Fein et al., 2019). The researchers had to heat up the molecules and fire them in a six and a half foot long beam. After accounting for gravity, heat, and other factors, the researchers were able to show that even these mega-size molecules could experience superposition, being in two places at the same time technically. Don't get too excited though! Science is nowhere near making doubles of complex systems like humans.

Breaking Reality

How do you define real? For all of us living every day in the macroscopic world, we rely on our five senses to help us navigate life. Are we hungry? Is this water safe to drink? Did the first car hit the second car? All of these situations require us to use our observation powers and logic together. However, as we now know, things aren't so easily determined on a microscopic level.

If Bell's inequality theorem could be broken in a lab, it would show that the Copenhagen school of thought was right—reality was determined by the observer. In 2019 at Heriot-Watt University in Edinburgh, Scotland, a small quantum computer was used to test the idea that observers could impact reality. Inside this computer, entangled photons were measured for weeks to gain statistics. One pair of entangled photons were the 'coins,' while another pair "tossed the coins." A final pair observed and measured the "coin toss." Over time, enough numbers were gathered to prove that, controlled by the observer, the result of the coin toss was all dependent on the observer alone.

A few issues within the experiment remain and not all possibilities are explained. This experiment did not prove that there are signals going faster than the speed of light. Neither could the experiment prove that the observers were free to choose the measurements they made. Can photons even count as observers? According to Časlav Brukner, yes. Therefore, while the experiment is not entirely complete, it does appear to show that in the quantum world, reality really does depend on your point of view!

Quantum Supremacy

How fast will these quantum computers be able to crunch numbers and run software? To get a better idea, check out Google's claim in 2019. A team at the University of California published a paper, which said that they had succeeded in building a real quantum computer.

You might be wondering how scientists and engineers can prove that a computer is quantum. The key lies in figuring out a calculation that only a quantum computer could handle quickly.

Classical supercomputers are famous for calculating very complex formulas, but there are a few special calculations that can't be handled easily, even by a supercomputer. According to Google, its quantum computer completed the calculation. For a classical supercomputer, this calculation would have taken around 10,000 years to figure out! Although this calculation doesn't have any real or practical usefulness, it stands as good proof that a computer actually has the power of quantum computing.

Moving forward, as Google works with NASA and as IBM investigates quantum computing, these and other companies around the world seek ways to not only successfully prove quantum computing has been achieved, but also to create prototypes that one day may show up in everyone's homes.

The Future of Research in Quantum Physics

Where is this path through the world of quantum physics going to take us? Who knows! It is hard to tell at this stage as advancements in technology exponentially increase, old questions are resolved, and new questions are asked. In 2021 alone, the first steps toward quantum internet were taken with the creation of a multi-node quantum network. At the same time, the AI, Theseus, began to use machine learning algorithms to solve calculations about complex quantum states. Even some of Einstein's proposals, like the possibility of creating matter out of light, has been proven possible.

A few decades ago, the idea that a particle could be created out of colliding light particles might have seemed crazy, but the world of quantum physics is always changing. Like the particles observed within the uncertain and mysterious quantum world, this field of study isn't afraid to ask difficult questions and push at the boundaries of what we define as reality. It makes you wonder what else will be accomplished going forward!

PART 2

CONCLUSION

The reality we can put into words is never reality itself.

–Werner Heisenberg

Our trip went well, didn't it? On our way home, we say goodbye to the thick forests and look forward to stopping in at our favorite coffee shop. Goodbye, wilderness. Hello, urban jungle! As you tap your bank card for the coffee drinks, you might pause and think about the simple transaction you are so used to doing day after day. How much of the quantum world and the study of quantum physics do you rely on?

Whether we are spending time in nature or enjoying the technological comforts of our home, it's hard to imagine that the environment we are experiencing is based on unpredictable wave particles. The medical treatments we rely on and the computers we use were built with a ground-breaking knowledge of quantum physics. Looking at our familiar faces in the mirror, it probably feels surreal to recognize that deep down, on a subatomic level, you consist of something as momentary as waves and energy. Not only that, but these wave particles and points of energy don't obey the same laws of nature of time and reality that we experience on a daily basis.

On some level, it is easy to view quantum physics as mysterious, strange, or too difficult to understand. After all, looking at the whiteboards filled with theoretical mathematical formulas, you may begin to recognize the amount of work and analysis that goes on behind the scenes. However, understanding the basics of quantum physics is more than just finding something new about 'wacky' science. It can also provide an opportunity for you to understand the necessary journey that all scientists must walk in the pursuit of truth. Scientific discoveries and theories are rarely flawless, and they often require long periods of work and change in order to define reality more accurately.

Whether their findings were what they wanted or expected, the pioneers of quantum physics attempted to accept and understand the strange new world they discovered. Over time, some theories and models appear to have been proven wrong, and others have yet to be confirmed or disproven. Perhaps, then, quantum physics can also give us a great lesson about the importance of curiosity, perseverance, and passion.

Understanding Quantum Physics

A lot of famous physicists and scientists are known for suggesting that if you think you understand quantum physics, you probably don't understand it at all. The main idea behind these kinds of statements is to emphasize the complexity, challenging nature, and cutting-edge uniqueness of this field of study. Today, we take these discoveries and observations for granted, but for Einstein and his fellow scientists, entanglement, wave particle duality, and superposition weren't so easy to accept. It felt like the world had gone topsy-turvy. All rules seemed to be flying out the window in the microscopic, subatomic realm of science.

Experience teaches us that when you throw a ball up in the air, it will come down. We know that we can't drink ice cubes like water. We understand that when we hold hands, we can let go at any time. We recognize that it is impossible to reverse everything in life, especially scrambled eggs and broken mugs. Life appears to rely on our interactions with the world through our five senses. Not only do we rely on experiences and observation, but we also subconsciously require stability, logic, and coherence, especially on larger social scales. These realities around us determine how we respond to our world and have driven our need for improved technology, better medication, and convenient living. Yet, when we look at the atomic world that builds the entire cosmos in complex chains of molecules, we find a different story.

In quantum physics, we discover that something can be two separate things at the same time. Wave particle duality seems to suggest that a particle is both wave and particle before it is observed. When it is observed, the particle appears to show up randomly within the parameters of the wave. Then there is superposition, where a particle not only may experience two states at the same time but may also share its spacetime with other particles. On top of that, quantum entanglement gives us "spooky action at a distance," as entangled particles appear to directly influence each other with different results depending on the measurements used. Finally, retrocausality could also be in play as the measurements of particles seem to affect their past as much as their present. All of this might seem difficult to prove, much less accept. Still, quantum physicists believe that with time and research, these mysteries will be more easily explained. Until then, the field of quantum physics offers a world of cutting-edge technology and scientific discoveries.

The Power of the Quantum

So, are we going to have faster smartphones? Are we going to have more efficient global banking networks? Will video games become more complex and fun? Will the Internet and cloud networks perform even better? Right now, it might be easy to say, "Yes!" However, realistically speaking, a lot of quantum physics research right now is focused on laying the foundation for quantum computers and other technological devices that will support more in-depth scientific research. As qubit chips increase in power, researchers will be able to complete larger and more complex calculations. This in turn might help push forward innovation in other areas as quantum mechanics is applied to practical applications.

Already, we are seeing how quantum physics has impacted our world. Whenever we swipe our cards, use barcodes, get an MRI scan, or use a laser pointer, we are enjoying the results of practically applied quantum physics research. Today, quantum clocks, codes, and microscopes are pushing the envelope in IT and research fields. With quantum clocks, our satellite transmissions are more efficient and stable, allowing for more accurate GPS signals to our phones—all thanks to the work of quantum physicists who have uncovered some of the secrets of the universe.

As for the future, there are still many great unknowns. A few quantum and theoretical physicists are still trying to figure out how gravity works with the other forces of nature on a subatomic level. At the same time, many other researchers are focused on improving existing inventions or creating new thought models from scratch in order to answer old questions that remain unanswered. As we saw in the final chapter, many exciting experiments are ongoing, answering some questions but also revealing more mysteries to explore. It is indeed exciting to think of what might come next!

Physics and You

The study of quantum physics requires an acceptance that few things can be objectively verified. It's challenging, but it allows for alternative ways to approach the universe around you. Is it overwhelming at first? Perhaps, but with perseverance and an openness to change, quantum physics has a lot to offer. After reading this book, you might want to tackle the theories and questions of quantum physics. If you find mathematics and physics interesting already, don't let anything hold you back!

When it comes to tackling quantum physics, a good place to start is in the past. Begin with classical notions of physics because quantum physics isn't just about what is going on right now but is also deeply tied to ideas posed in the past. By understanding the history of scientific inquiry, you will better understand why certain conclusions are agreed on today. Starting with Sir Isaac Newton, you can explore the writings and work of various physicists mentioned in Chapter 1: Albert Einstein, Niels Bohr, Erwin Schrödinger, Werner Heisenberg, and David Bohm. You might also want to check out the work of Louis de Broglie, Paul Dirac, Richard Feynman, Julian Schwinger, and Sin-Itiro Tomonaga. In particular, Richard Feynman's famous series, *The Feynman Lectures on Physics*, might be a great place to start. More up-to-date discoveries and papers are shared on *arXiv.org*, which you can access online.

Whether you want to dive deep into quantum physics, make it your career, or just enjoy keeping up to date with new discoveries, you will have to be prepared for challenge, uncertainty, and some "spooky action." At the same time, however, you might find yourself more appreciative of the amazingly complex cosmos in which we live. With each new, strange discovery, the quantum realm will throw down the gauntlet. The subjective nature of quantum physics might make you feel like understanding the truth will never be achieved.

Don't give up. Listen closely to what the quantum realm is saying to you: Nothing is what it seems, but the potential of reality is endless. Accepting change and errors is part of an important process for growth and learning, keep moving forward and transform. Perhaps, in the end, the journey is more important than the destination.

INCLUDED

For additional resources and free downloads that are included with your book purchase, visit **www.pantheonspace.com/Qplus**

These include classic works like:

- ***The Theory of Relativity* by Albert Einstein**
- **The Origin of Quantum Theory by Max Planck**
- **Eight Lectures delivered at Columbia by Max Planck**
- **The Nature of the Physical World by Sir Arthur Stanley Eddington**

You can also download my extended glossary, which offers in-depth explanations and hundreds of terms to deepen your understanding of quantum physics. All of these digital files, along with even more exclusive content, come with this second edition of the book. I encourage you to explore these materials and enhance your learning experience right away!

Be notified of new releases by following us at **www.amazon.com/author/xyz**

Keep the conversation going by joining us on FB **www.facebook.com/pantheonspace**

GLOSSARY

101 TERMS DEFINED

A

Absolute zero: The lowest possible temperature, where particles have minimal motion.

Amplitude: The maximum extent of a wave or oscillation, measured from its resting point. In quantum mechanics, amplitude is related to the probability of finding a particle in a particular state.

Antimatter: Matter composed of antiparticles, which have the same mass but opposite charges compared to normal matter. When matter and antimatter meet, they annihilate each other, releasing energy.

Atom: The basic unit of matter, consisting of protons, neutrons, and electrons.

B

Baryons: A type of subatomic particle made of three quarks, including protons and neutrons. Baryons are part of the hadron family and make up most of the visible matter in the universe.

Bell's Theorem: A principle that shows certain predictions of quantum mechanics are incompatible with local hidden variables.

Blackbody radiation: Electromagnetic radiation emitted by an idealized object that absorbs all radiation incident upon it. The spectrum of blackbody radiation depends only on the temperature of the object.

Bloch spheres: A geometric representation used in quantum mechanics to describe the state of a two-level quantum system, such as a qubit in quantum computing.

Bohr Model: A model of the atom where electrons orbit the nucleus in specific energy levels.

Bohr radius: The most probable distance between the nucleus and the electron in a hydrogen atom.

Boson: A type of particle that follows Bose-Einstein statistics and can occupy the same quantum state as other bosons. Examples include photons and the Higgs boson.

Bra-ket notation: A mathematical notation used in quantum mechanics to describe quantum states.

C

Casimir Effect: A quantum phenomenon where two uncharged metal plates in a vacuum can attract each other due to quantum fluctuations.

Collapse: The reduction of a quantum wave function to a single state upon measurement.

Coherence: The property of a quantum system where all parts of the wave function are in sync.

Commutator: An operator that shows the relationship between two quantum operators and their order.

Complementarity: Bohr's principle that objects can display particle-like or wave-like behavior, depending on the experimental setup.

Compton Scattering: The increase in wavelength (and decrease in energy) of X-rays or gamma rays when they scatter off electrons.

Conservation of energy: A principle stating that energy cannot be created or destroyed, only transformed.

Copenhagen Interpretation: A common explanation of quantum mechanics that suggests particles exist in multiple states until observed.

Coulomb forces: The electrostatic forces between charged particles, described by Coulomb's law, where opposite charges attract and like charges repel.

D

de Broglie Wavelength: The wavelength associated with any moving particle, illustrating wave-particle duality.

Decoherence: The process by which a quantum system loses its quantum behavior, typically due to interaction with the environment.

Degeneracy: The condition where two or more quantum states have the same energy.

Diffraction: The bending of waves around obstacles or the spreading of waves as they pass through a narrow opening. Diffraction occurs with all types of waves, including light and sound.

Dirac equation: A relativistic equation describing the behavior of fermions, such as electrons.

Double-slit experiment: A notable experiment in quantum mechanics showing wave-particle duality through interference patterns.

E

Ehrenfest theorem: A result connecting classical mechanics with quantum mechanics.

Eigenstate: A quantum state corresponding to a specific eigenvalue.

Eigenvalue: A possible value of an observable in a quantum system.

Electron: A subatomic particle with a negative charge that orbits the nucleus of an atom. Electrons play a key role in chemical reactions and electricity.

Electron cloud: A region around the nucleus where an electron is likely to be found.

Electron spin: A quantum property of electrons that gives rise to magnetic moments.

Energy Levels: Discrete values of energy that an electron in an atom can have.

Entanglement: A quantum phenomenon where two or more particles become linked so that the state of one particle directly influences the state of another, no matter the distance.

Entropic Time Arrow: The idea that time moves in the direction of increasing entropy in a quantum system.

Excited State: When an electron is in a higher energy level than its ground state.

Exclusion principle: Pauli's principle states that no two fermions can simultaneously occupy the same quantum state.

Expectation value: The average value of an observable for a quantum system in a given state.

F

Field: A physical quantity, represented by numbers at each point in space and time, that can describe forces like gravity or electromagnetism.

Fermion: A type of particle that follows Fermi-Dirac statistics, such as electrons and protons, which obey the Pauli exclusion principle.

Feynman diagram: A visual representation of particle interactions in quantum field theory.

Feynman path integral: A formulation of quantum mechanics that sums over all possible paths a particle can take.

Fine structure constant: A fundamental physical constant characterizing the strength of electromagnetic interactions.

Fission: The process in which the nucleus of an atom splits into two or more smaller nuclei, often releasing energy. Fission is the basis of nuclear power and atomic bombs.

Frequency: The number of oscillations or cycles a wave completes in one second, typically measured in Hertz (Hz). Frequency is inversely related to wavelength.

Fusion: The process by which two atomic nuclei combine to form a heavier nucleus, releasing energy. Fusion powers stars, including our Sun.

G

General relativity: Albert Einstein's theory that describes gravity as the curvature of spacetime caused by mass and energy. It explains large-scale phenomena like black holes and the expansion of the universe.

Gluons: Particles that act as the force carriers for the strong nuclear force, binding quarks together to form protons, neutrons, and other hadrons.

Ground State: The lowest energy state of an atom or particle.

H

Hadrons: Particles made up of quarks, held together by the strong nuclear force. Baryons (like protons and neutrons) and mesons are examples of hadrons.

Half-life: The time it takes for half of the atoms in a sample of a radioactive substance to decay. Each radioactive isotope has a specific half-life.

Heisenberg uncertainty principle: The principle states that the more precisely you know a particle's position, the less precisely you can know its momentum, and vice versa.

Higgs Boson: The particle associated with the Higgs field, responsible for giving mass to other particles. Its discovery in 2012 confirmed a key aspect of the Standard Model of particle physics.

Higgs field: A field that permeates space and gives mass to elementary particles through their interaction with it. The Higgs boson is the associated particle of this field.

Hilbert space: A mathematical space used to represent quantum states.

I

Infrared: Electromagnetic radiation with wavelengths longer than visible light but shorter than microwaves. Infrared light is commonly associated with heat.

Interference: A quantum phenomenon where waves overlap, reinforcing or canceling each other.

Ionization: The process of removing an electron from an atom or molecule.

Ions: Atoms or molecules that have gained or lost one or more electrons, giving them a positive or negative charge.

Isotope: Variants of a chemical element that have different numbers of neutrons.

K

Ket: A vector in Hilbert space representing a quantum state.

Kinetic energy: The energy an object possesses due to its motion. In quantum mechanics, kinetic energy plays a role in the behavior of particles at the atomic and subatomic levels.

L

Laser: A device that emits coherent light through the process of optical amplification.

Lattice: A regular arrangement of points in space, often used to describe crystal structures in quantum systems.

Leptons: A family of elementary particles that includes electrons, muons, and neutrinos. Unlike quarks, leptons do not experience the strong nuclear force.

Light: Electromagnetic radiation visible to the human eye, spanning a small portion of the electromagnetic spectrum. Photons are the fundamental particles of light.

M

Measurement problem: The unresolved issue of how and when a quantum system collapses into a definite state.

M-theory: A proposed unified theory in theoretical physics that attempts to describe the fundamental forces and particles of the universe through 11-dimensional spacetime.

Magnetic: Referring to the property of materials or particles that experience a force when exposed to a magnetic field. Magnetic forces arise from the motion of charged particles.

Mass: The amount of matter in an object, which determines its resistance to acceleration and its gravitational attraction to other objects.

Matter: Anything that has mass and occupies space. Matter exists in various forms, such as solids, liquids, gases, and plasma.

Muons: Elementary particles similar to electrons but with a much greater mass. Muons are unstable and decay into electrons, neutrinos, and other particles.

N

Neutrino: A nearly massless, neutral particle that interacts weakly with matter.

Neutron: A neutral particle found in the nucleus of an atom.

Nuclei: The positively charged center of an atom, made up of protons and neutrons. The nucleus contains most of the atom's mass.

Nucleons: The particles (protons and neutrons) that make up an atomic nucleus.

Nucleus: The positively charged center of an atom, containing protons and neutrons.

O

Observable: A physical quantity in a quantum system that can be measured, such as position, momentum, or spin.

Olfaction: The sense of smell, not typically associated with quantum physics but studied in the context of quantum biology to explain how quantum processes may influence biological functions like detecting odors.

Operator: A mathematical entity that acts on wave functions to represent observables.

Oscillations: Regular variations or movements around a central point, such as the back-and-forth motion of a pendulum or the repetitive waves in quantum systems.

P

Particle-wave duality: The concept that quantum entities exhibit both particle and wave characteristics.

Pauli Exclusion Principle: A rule that no two fermions (like electrons) can occupy the same quantum state at the same time.

Photoelectric effect: The emission of electrons from a metal surface when exposed to light, explained by Einstein in terms of photons.

Photon: The fundamental particle of light that carries electromagnetic force.

Photon Energy: The energy carried by a photon, proportional to its frequency.

Pions: A type of meson composed of one quark and one antiquark, which mediates the strong force between nucleons in an atomic nucleus.

Pixels: The smallest unit of a digital image or display, representing the intensity of light or color in a visual system.

Planck's Constant: A fundamental constant **(h)** that relates the energy of a photon to its frequency, central to quantum mechanics.

Planck Energy: The energy scale at which quantum gravitational effects are expected to become important.

Planck's law: Describes the electromagnetic radiation emitted by a blackbody.

Planck Length: The smallest meaningful unit of length in quantum mechanics and general relativity.

Planck Temperature: The highest temperature before quantum effects dominate all forces.

Planck Time: The smallest meaningful unit of time in quantum mechanics.

Polarization: The orientation of the oscillations of a wave, especially light waves.

Positron: The antiparticle of the electron, with the same mass but opposite charge.

Potential well: A concept describing the trapping of a particle in a region of lower potential energy.

Prime factors: Numbers that are prime and multiply together to give another number. In quantum computing, prime factorization plays a key role in cryptography.

Proton: A positively charged particle in the nucleus of an atom.

Probability Amplitude: The value related to the likelihood of a particular quantum event occurring, often represented by the wave function.

Probability density: The likelihood of finding a particle in a given position.

Probability Wave: A mathematical description that predicts the likelihood of finding a particle in a particular position.

Q

Quantization: The concept that certain physical properties can only exist in discrete values.

Quantum: The smallest possible discrete unit of any physical property, like energy or matter.

Quantum Chromodynamics (QCD): The theory that describes the behavior of quarks and gluons, which make up protons and neutrons.

Quantum Circuit: A model for quantum computation where operations are performed on qubits using quantum gates.

Quantum computer: A type of computer that uses quantum bits (qubits) and the principles of superposition and entanglement to perform calculations.

Quantum cryptography: The use of quantum mechanics to encrypt and transmit data securely.

Quantum Electrodynamics (QED): A theory describing how light and matter interact using quantum mechanics and the theory of special relativity.

Quantum Error Correction: Techniques to protect quantum information from errors due to decoherence and other quantum noise.

Quantum Field: A field that exists throughout space and time, associated with quantum particles.

Quantum Field Theory: A framework for describing the behavior of quantum fields, combining quantum mechanics and special relativity.

Quantum Fluctuations: Temporary changes in energy that occur spontaneously in a quantum system.

Quantum foam: A concept suggesting that spacetime is not smooth but fluctuates at the quantum level.

Quantum Fourier Transform: A mathematical operation critical for certain quantum algorithms, such as Shor's algorithm.

Quantum gravity: A theoretical framework attempting to describe gravity according to the principles of quantum mechanics.

Quantum Harmonic Oscillator: A model describing particles in a potential well that oscillates about an equilibrium position.

Quantum Interference: The phenomenon where the probability of quantum outcomes is affected by wave-like properties, leading to constructive or destructive interference.

Quantum Key Distribution (QKD): A method to securely share encryption keys using quantum principles.

Quantum Leap (Quantum Jump): A sudden change of an electron from one energy level to another in an atom.

Quantum logic gate: A basic unit of operation in quantum computing, which manipulates qubits.

Quantum Mechanics: The branch of physics that deals with the behavior of particles on an atomic and subatomic scale.

Quantum Metrology: The use of quantum mechanics to improve precision measurements beyond classical limits.

Quantum number: A number that describes the properties of particles in a quantum system.

Quantum State: The set of information that fully describes a quantum system, like the position or energy of a particle. Typically defined by a wave function.

Quantum superposition: The principle that a quantum system can exist in multiple states simultaneously.

Quantum Supremacy: The point where a quantum computer can solve problems faster than classical computers.

Quantum Well: A potential energy structure that confines particles, often used in semiconductor devices.

Quantum Zeno Effect: A phenomenon where a quantum system's evolution is hindered by frequent measurement, seemingly "freezing" its state.

Quark: A fundamental particle that makes up protons and neutrons, each with a different mass and charge, known as flavors (up, down, charm, strange, top, bottom).

Qubit: The basic unit of information in a quantum computer.

R

Radio waves: Electromagnetic waves with the longest wavelengths and lowest frequencies, commonly used for communication.

Radioactivity: The spontaneous emission of radiation from unstable atomic nuclei, releasing particles and energy in the form of alpha, beta, or gamma radiation.

Refraction: The bending of light or other waves when passing from one medium to another, such as air to water, causing a change in speed and direction.

Relativity (General and Special): The theory developed by Einstein that explains how time, space, and gravity interact on large scales.

Renormalization: A process in quantum field theory to remove infinities and make predictions finite and meaningful.

Resonance: The condition in which a quantum system absorbs energy efficiently from an external source.

Rydberg Constant: A constant related to the energy of electron transitions in hydrogen.

S

Scalar bosons: Particles that carry forces in quantum fields but are not associated with any spin. The Higgs boson is a type of scalar boson.

Schrödinger Equation: A key equation in quantum mechanics that describes how the wavefunction of a system evolves over time.

Schrödinger's Cat: A thought experiment illustrating the idea of superposition and measurement in quantum mechanics.

Schwarzschild radius: The radius at which the escape velocity from a mass equals the speed of light, associated with black holes.

Singularity: A point in space where density and gravity become infinite, often found in black holes.

Spin: A quantum property of particles that describes their intrinsic angular momentum.

Spin-Orbit Coupling: The interaction between a particle's spin and its motion, affecting energy levels.

Squared: A mathematical operation where a number or quantity is multiplied by itself. In quantum mechanics, probabilities are often found by squaring the amplitude of a wave function.

SQUIDs: Superconducting Quantum Interference Devices, which are highly sensitive instruments used to measure extremely subtle magnetic fields.

Standard Model: The theory describing the electromagnetic, weak, and strong nuclear forces.

State: The condition of a quantum system, defined by properties like energy or position.

String theory: A theoretical framework in which particles are one-dimensional strings instead of point-like.

Strong force: The force that holds protons and neutrons together in an atomic nucleus.

Superconductivity: A quantum phenomenon where certain materials conduct electricity with zero resistance at low temperatures.

Superconductor: A material that can conduct electricity with zero resistance at extremely low temperatures.

Superposition: The ability of a quantum system to exist in multiple states at once.

T

Tachyon: A hypothetical particle that always moves faster than the speed of light.

Teleportation (Quantum): A process where quantum information is transmitted between distant locations using entanglement.

Trapped ions: Ions that are confined using electromagnetic fields, often used in quantum computing to perform precise quantum operations.

Tunneling: The quantum phenomenon where a particle passes through a barrier it classically couldn't surmount.

U

Ultraviolet (UV): Electromagnetic radiation with wavelengths shorter than visible light but longer than X-rays, often responsible for causing sunburns.

Uncertainty principle: The idea that you cannot know both a particle's position and momentum with perfect accuracy at the same time.

Unitarity: A principle in quantum mechanics that ensures the total probability of all possible outcomes of an event is 1.

V

Vacuum energy: The actual energy value associated with the vacuum state, arising from quantum fluctuations and considered as a background energy permeating the universe.

Vacuum State: Represents the "empty" space in quantum field theory, where all fields are in their lowest energy state, often associated with the concept of "zero-point energy.

W

Wavefunction: A mathematical function that describes the quantum state of a system, providing probabilities for where a particle might be found.

Wavelength: The distance between successive crests of a wave, determining the wave's energy and frequency.

Wave-particle duality: The theory that all particles exhibit both wave-like and particle-like properties.

W bosons: Particles that mediate the weak nuclear force, responsible for processes like beta decay in radioactive materials.

Weak Nuclear Force: One of the four fundamental forces of nature, responsible for radioactive decay.

X

X-rays: A form of electromagnetic radiation with wavelengths shorter than ultraviolet light, commonly used in medical imaging and for studying the structure of materials.

Z

Z bosons: Particles that, along with W bosons, mediate the weak nuclear force, one of the fundamental forces in nature.

Zeptosecond: One trillionth of a billionth of a second (10^-21 seconds), a very short unit of time used in high-precision measurements in physics.

Zero-point energy: The lowest possible energy a quantum system can have, even at absolute zero.

Zeeman Effect: The splitting of spectral lines in an atom due to the presence of a magnetic field.

Zone plate: An optical device used to focus light via diffraction, often used in X-ray optics.

Zeno effect: A quantum phenomenon where frequent observation of a system prevents its evolution.

Zitterbewegung (Trembling motion): The rapid oscillatory motion predicted for relativistic particles like electrons.

Zitterbewegung Oscillation: The rapid oscillatory motion predicted for relativistic particles like electrons.

REFERENCES

SECOND EDITION, PART 1

2023 | the-humble-essayist. (n.d.). The-humble-essayist. https://www.the-humble-essayist.com/2023

A Dictionary of Physics. (2015). In *Oxford University Press eBooks.* https://doi.org/10.1093/acref/9780198714743.001.0001

Admin, C. (2003, September 12). *Mr. Uncertainty: Part 2: The battle over Heisenberg.* Christianity Today. https://www.christianitytoday.com/2000/05/mr-uncertainty-part-2-battle-over-heisenberg/

Admin, J. (2023, October 8). Bohr's Exile And The Atomic Bomb. *JYP.* https://www.journalofyoungphysicists.org/post/bohr-s-exile-and-the-atomic-bomb

Albert Einstein: Overcame Early School Challenges, Won Nobel Prize | Aug 27, 2024. (n.d.). https://www.elephantlearning.com/post/albert-einstein-overcame-school-challenges-won-nobel-prize

Author, N. (2024a, August 16). *Werner Heisenberg: controversial scientist – Physics World.* Physics World. https://physicsworld.com/a/werner-heisenberg-controversial-scientist/

Author, N. (2024b, August 16). *Werner Heisenberg: controversial scientist – Physics World.* Physics World. https://physicsworld.com/a/werner-heisenberg-controversial-scientist/

Bohmian Mechanics (Stanford Encyclopedia of Philosophy). (2021, June 14). https://plato.stanford.edu/entries/qm-bohm/

Cotton, D. (2023, January 19). *Albert Einstein Costume - Fancy dress Ideas.* Costume Rocket. https://costumerocket.com/albert-einstein-costume/

Crease, R. P. (2024, March 14). *When Bose wrote to Einstein: the power of diverse thinking – Physics World.* Physics World. https://physicsworld.com/a/when-bose-wrote-to-einstein-the-power-of-diverse-thinking/

Cushing, J. T. (1991). Quantum Theory and Explanatory Discourse: Endgame for Understanding? *Philosophy of Science, 58*(3), 337–358. https://www.jstor.org/stable/187936

De Los Angeles Fanaro, M., Arlego, M. J. F., Otero, M. R., & Elgue, M. (2018). Students' Interpretations of Quantum Mechanics Concepts from Feynman's Sum of all Paths Applied to Light. *International Journal of Physics & Chemistry Education, 10*(2), 41–47. https://doi.org/10.51724/ijpce.v10i2.19

Expert. (2024, September 8). *"Paul Dirac: Antimatter's Predictor – Merging Mathematics and Physics"* Editverse. https://editverse.com/paul-dirac-quantum-mechanics-antimatter/

Gardner, H. (2021, March 29). *An Extraordinary Commentary on the Festschrift "Mind, Work, and Life" — Howard Gardner.* Howard Gardner. https://www.howardgardner.com/howards-blog/an-extraordinary-commentary-on-the-festschrift-mind-work-and-life

Gibney, E. (2024, June 4). Exotic Quantum 'Bose-Einstein Condensate' State Finally Achieved with Molecules. *Scientific American.* https://www.scientificamerican.com/article/exotic-quantum-bose-einstein-condensate-state-finally-achieved-with/

Imdad.Alvi, & Imdad.Alvi. (2022, January 4). *Black body radiation and planck's radiation law.* Ox Science. https://oxscience.com/black-body-radiation/

Kaku, M. (2024, October 3). *Albert Einstein | Biography, Education, Discoveries, & Facts.* Encyclopedia Britannica. https://www.britannica.com/biography/Albert-Einstein/From-graduation-to-the-miracle-year-of-scientific-theories

Kazemian, S., & Fanchini, G. (2024). Influence of higher-order electron-phonon interaction on the electron-related lattice thermal properties of two-dimensional Dirac crystals. *Physical Review. B./Physical Review. B, 109*(20). https://doi.org/10.1103/physrevb.109.205422

Kbconlinegame, K. (2024, January 12). *Indian Nobel Laureates GK Questions and Answers with PDF download.* https://kbconlinegame.com/indian-nobel-laureates-gk-questions-answers/

Khan, S. (2023, September 14). *Manhattan Project Scientists Believed the Way We Get Out Alive is World Government.* Inkstick. https://inkstickmedia.com/manhattan-project-scientists-believed-the-way-we-get-out-alive-is-world-government/

Muzdakis, M. (2021, February 3). *29 legendary scientists came together in the "Most intelligent photo" ever taken.* My Modern Met. https://mymodernmet.com/the-solvay-conference-photo/

Nolte, D. D. (2023, October 24). *Dirac: From Quantum Field Theory to Antimatter.* Galileo Unbound. https://galileo-unbound.blog/2019/02/23/dirac-quantum-field-theory-to-antimatter/

Olwell, R. (1999). Physical Isolation and Marginalization in Physics: David Bohm's Cold War Exile. *Isis*, *90*(4), 738–756. https://www.jstor.org/stable/237658

Ornes, S. (2017a). How Bose–Einstein condensates keep revealing weird physics. *Proceedings of the National Academy of Sciences*, *114*(23), 5766–5768. https://doi.org/10.1073/pnas.1707804114

Ornes, S. (2017b). How Bose–Einstein condensates keep revealing weird physics. *Proceedings of the National Academy of Sciences*, *114*(23), 5766–5768. https://doi.org/10.1073/pnas.1707804114

Parr, F. (2024, September 9). *23 April: On this day in history.* HistoryExtra. https://www.historyextra.com/on-this-day/23-april-on-this-day-in-history/

Physics History January 2005. (n.d.). American Physical Society. https://www.aps.org/apsnews/2005/01/einstein-photoelectric-effect

Popova, M. (2017a, November 7). *Love After Life: Nobel-Winning Physicist Richard Feynman's Extraordinary Letter to His Departed Wife.* The Marginalian. https://www.themarginalian.org/2017/10/17/richard-feynman-arline-letter/

Popova, M. (2017b, November 7). *Love After Life: Nobel-Winning Physicist Richard Feynman's Extraordinary Letter to His Departed Wife.* The Marginalian. https://www.themarginalian.org/2017/10/17/richard-feynman-arline-letter/

Sahni, V. (2022). Perspectives on determinism in quantum mechanics: Born, Bohm, and the "Quantal Newtonian" laws. *The Journal of Chemical Physics, 157*(24). https://doi.org/10.1063/5.0130945

Sevush, S. (2016). Quantum Consciousness. In *Springer eBooks* (pp. 231–249). https://doi.org/10.1007/978-3-319-33708-1_11

S.N. Bose Scholar Exchange | WINStep Forward. (n.d.). WINStep Forward. https://www.winstepforward.org/sn-bose-scholar-exchange/

Soul, Q., & Soul, Q. (2023, December 26). Max Planck's impact on the birth of quantum physics. *ScienceShot - Professional Insights, Unique Perspectives.* https://www.scienceshot.com/post/max-plancks-impact-on-the-birth-of-quantum-physics

Szabó, L. E. (2010). What remains of probability? In *Springer eBooks* (pp. 373–379). https://doi.org/10.1007/978-90-481-9115-4_26

The dual nature of light as reflected in the Nobel archives. (n.d.-a). NobelPrize.org. https://www.nobelprize.org/prizes/themes/the-dual-nature-of-light-as-reflected-in-the-nobel-archives/

The dual nature of light as reflected in the Nobel archives. (n.d.-b). NobelPrize.org. https://www.nobelprize.org/prizes/themes/the-dual-nature-of-light-as-reflected-in-the-nobel-archives/

The Nobel Prize in Physics 1918. (n.d.-a). NobelPrize.org. https://www.nobelprize.org/prizes/physics/1918/planck/facts/

The Nobel Prize in Physics 1918. (n.d.-b). NobelPrize.org. https://www.nobelprize.org/prizes/physics/1918/planck/facts/

The Nobel Prize in Physics 1921. (n.d.-a). NobelPrize.org. https://www.nobelprize.org/prizes/physics/1921/einstein/lecture/

The Nobel Prize in Physics 1921. (n.d.-b). NobelPrize.org. https://www.nobelprize.org/prizes/physics/1921/einstein/facts/

The Nobel Prize in Physics 1921. (n.d.-c). NobelPrize.org. https://www.nobelprize.org/prizes/physics/1921/summary/

The Nobel Prize in Physics 1921. (n.d.-d). NobelPrize.org. https://www.nobelprize.org/prizes/physics/1921/einstein/biographical/

Woit. (n.d.). *Not So Spooky Action at a Distance | Not Even Wrong.* https://www.math.columbia.edu/~woit/wordpress/?p=11056&cpage=1

Youvan, D. C. (2024a). Convergence of Thought: Exploring the Influence of Maya and Advaita Vedanta on Erwin Schrödinger's Quantum. . . *ResearchGate.* https://doi.org/10.13140/RG.2.2.14652.21122

Youvan, D. C. (2024b). Harassment in the McCarthy Era: The Impact on Notable Individuals from Various Fields. *ResearchGate.* https://doi.org/10.13140/RG.2.2.15200.49924

REFERENCES

SECOND EDITION, PART 2

American Institute of Physics. (2021). *Bright idea: The first lasers.* Bright Idea. https://history.aip.org/exhibits/laser/index.html

American Museum of Natural History. (2019). *Quantum theory.* American Museum of Natural History. https://www.amnh.org/exhibitions/einstein/legacy/quantum-theory

Ananthaswamy, A. (2021, July 14). *AI designs quantum physics experiments beyond what any human has conceived.* Live Science. https://www.livescience.com/ai-designs-quantum-physics-experiments.html

Aron, J. (2013). *Quantum chip connected to internet is yours to command.* New Scientist. https://www.newscientist.com/article/dn24159-quantum-chip-connected-to-internet-is-yours-to-command/?ignored=irrelevant

Arvin Ash. (2020). *The EPR Paradox & Bell's inequality explained simply.* YouTube. https://www.youtube.com/watch?v=f72whGQ31Wg

Aspect, A. (2007). To be or not to be local. *Nature, 446*(7138), 866–867. https://doi.org/10.1038/446866a

Assaraf, J. (2010, August 17). *Why you should be aware of quantum physics.* John Assaraf: Achieve Even More.

https://johnassaraf.com/why-you-should-be-aware-of-quantum-physics-2/

Baker, L. (2020, February 6). *A Skeptic's take on the law of attraction.* Medium. https://medium.com/assemblage/a-skeptics-take-on-the-law-of-attraction-54046f5a23b#:~:text=In%20its%20most%20basic%20form%2C%20the%20Law%20of%20Attraction

Ball, P. (2018). *Quantum physics may be even spookier than you think.* Scientific American. https://www.scientificamerican.com/article/quantum-physics-may-be-even-spookier-than-you-think/

BestOfScience. (2009, December 24). *A brief history of quantum mechanics.* YouTube. https://www.youtube.com/watch?v=B7pACq_xWyw

Bohr, N. (1987). *Essays 1932-1957 on atomic physics and human knowledge.* Ox Bow Press.

Boyd, R. (2015, June 10). *Quantum physics neuroscience.* Energetics Institute. https://www.energeticsinstitute.com.au/articles/quantum-physics-neuroscience/

Boyle, A. (2008). *Discovery or doom? Collider stirs debate.* NBC News. https://www.nbcnews.com/id/wbna24556999

Brubaker, B. (2021, July 20). *How Bell's theorem proved "spooky action at a distance" is real.* Quanta Magazine.

https://www.quantamagazine.org/how-bells-theorem-proved-spooky-action-at-a-distance-is-real-20210720/

Brukner, C. (2018, March 28). Causality in a quantum world. *Physics Today.* https://physicstoday.scitation.org/do/10.1063/pt.6.1.20180328a/full/

Buttar, S. (2021, April 23). *On his 162nd anniversary of birth, here are 15 most beautiful quotes by Max Planck.* The Secrets of the Universe. https://www.secretsofuniverse.in/15-max-planck-quotes/

Carroll, S. (2019, September 7). Even physicists don't understand quantum mechanics. *The New York Times.* https://www.nytimes.com/2019/09/07/opinion/sunday/quantum-physics.html#:~:text=%E2%80%9CI%20think%20I%20can%20safely

Cena, C., & Foster, E. (2019). *The Theory of Relativity lesson for kids.* Study. https://study.com/academy/lesson/the-theory-of-relativity-lesson-for-kids.html

CERN. (2000). *The Higgs boson.* Home.cern. https://home.cern/science/physics/higgs-boson

Cho, A. (2010). The first quantum machine. *Science, 330*(6011), 1604–1604. https://doi.org/10.1126/science.330.6011.1604

Code.org. (2018a). *How computers work: Circuits and logic.* YouTube. https://www.youtube.com/watch?v=ZoqMiFKspAA

Code.org. (2018b). *How computers work: What makes a computer, a computer?* YouTube. https://www.youtube.com/watch?v=mCq8-xTH7jA

Coolman, R. (2014, September 26). *What is quantum mechanics?* Live Science; https://www.livescience.com/33816-quantum-mechanics-explanation.html

Corbin, K. (2021). *The Law of Attraction and quantum physics.* Law of Attraction Resource Guide. https://www.lawofattractionresourceguide.com/the-law-of-attraction-and-quantum-physics/

Coulter, D. (2011). *A freaky fluid inside Jupiter?* NASA Science. https://science.nasa.gov/science-news/science-at-nasa/2011/09aug_juno3

Crane, L. (2019). *Heat can quantum leap across a totally empty vacuum.* New Scientist. https://www.newscientist.com/article/2226783-heat-can-quantum-leap-across-a-totally-empty-vacuum/

Desgreniers, S. (2020). A milestone in the hunt for metallic hydrogen. *Nature, 577*(7792), 626–627. https://doi.org/10.1038/d41586-020-00149-7

Digital Watch Central. (2021). *How do digital watches keep time?* Digital Watch Central. https://digitalwatchcentral.com/how-do-digital-watches-keep-time/

Domenech, F. (2020, February 21). *Metallic hydrogen and the race to bag the unicorn of physics*. OpenMind. https://www.bbvaopenmind.com/en/science/physics/metallic-hydrogen-a-high-pressure-race-to-bag-the-unicorn-of-physics/

Einstein, A., & Born, M. (2014). *Born-einstein letters 1916-1955: Friendship, politics and physics in uncertain times.* Palgrave Macmillan.

Elhashash, M. (2020). *Modern physics: Its history, theories, and the practical experience of its virtual labs*. Praxilabs. https://blog.praxilabs.com/2020/07/30/modern-physics-history-theories/#The_Birth_of_Modern_Physics

Estes, D. (2020, April 3). *Physics of the law of attraction observe.* InnerSense. https://innersense-inc.com/physics-of-the-law-of-attraction-i-observe-therefore-i-am/

Everett, F. (2006, December). *James Clerk Maxwell: A force for physics*. Physics World. https://physicsworld.com/a/james-clerk-maxwell-a-force-for-physics/

February 2017, C. C. 13. (2017, February 13). *600-year-old starlight bolsters Einstein's "spooky action at a distance."* Space. https://www.space.com/35676-einstein-spooky-action-starlight-quantum-entanglement.html#:~:text=Entanglement%20is%20what%20Einstein%20referred

Fedrizzi, A., & Proietti, M. (2019). *Objective reality doesn't exist, quantum experiment shows*. Live Science.

https://www.livescience.com/objective-reality-not-exist-quantum-physicists.html

Fein, Y.Y., Geyer, P., Zwick, P., Kilaka, F., Pedalino, S., Mayor, M., Gerlich, S., Arndt, M. Quantum superposition of molecules beyond 25 kDa. *Nat. Phys.* 15, 1242–1245 (2019). https://doi.org/10.1038/s41567-019-0663-9

Filmer, J. (2013, December 25). *Sir Isaac Newton: Father of modern science.* Futurism. https://futurism.com/sir-isaac-newton-father-of-modern-science-2

Fore, M. (2019, March 15). *Physicists reverse time for tiny particles inside a quantum computer.* Live Science. https://www.livescience.com/65000-quantum-computer-turns-back-time.html

Gallardo, F. (2019, July 24). *The arrow of time.* Medium. https://medium.com/tourism-futures/the-arrow-of-time-644c32fdba26

Gell-Mann, M. (n.d.). *Murray Gell-Mann quotes.* BrainyQuote. https://www.brainyquote.com/quotes/murray_gellmann_305942

Gibney, E. (2019). Hello quantum world! Google publishes landmark quantum supremacy claim. *Nature, 574*(7779), 461–462. https://doi.org/10.1038/d41586-019-03213-z

Gouldson, S. (2014, June 2). *Untangling quantum entanglement: A child-friendly explanation.* JUMP! MAG. http://jumpmag.co.uk/quantum-entanglement/

Heisenberg, W. (1975). *Across the frontiers.* Harper & Row.

Heisenberg, W. (2020). *Great quotes by Werner Heisenberg.* Quotes. https://quotes.thefamouspeople.com/werner-heisenberg-5202.php

Herman, R. (2018). *How fast is the earth moving?* Scientific American. https://www.scientificamerican.com/article/how-fast-is-the-earth-mov/#:~:text=As%20schoolchildren%2C%20we%20learn%20that

History.com Editors. (2018, December 14). *The birth of quantum theory.* History. https://www.history.com/this-day-in-history/the-birth-of-quantum-theory

Hossenfelder, S. (2019). *How we know that Einstein's General Relativity can't be quite right.* YouTube. https://www.youtube.com/watch?v=Ov98y_DCvRY

Jenner, N. (2014, December). *Five practical uses for "spooky" quantum mechanics.* Smithsonian Magazine; https://www.smithsonianmag.com/science-nature/five-practical-uses-spooky-quantum-mechanics-180953494/

Jennings, R. (2020, October 23). *Manifesting is the new astrology.* Vox. https://www.vox.com/the-goods/21524975/manifesting-does-it-really-work-meme

Joint Quantum Institute. (2016). *Quantum superposition.* JQI. https://jqi.umd.edu/glossary/quantum-superposition

Kids Encyclopedia Facts. (2021a). *General relativity facts for kids.* Kiddle. https://kids.kiddle.co/General_relativity

Kids Encyclopedia Facts. (2021b). *Quantum entanglement facts for kids*. Kiddle. https://kids.kiddle.co/Quantum_entanglement

Kids Encyclopedia Facts. (2021c). *Wave–particle duality facts for kids*. Kiddle.
https://kids.kiddle.co/Wave%E2%80%93particle_duality

Kleppner, D. (2000). One hundred years of quantum physics. *Science, 289*(5481), 893–898.
https://doi.org/10.1126/science.289.5481.893

Kramer, M. (2013, August 14). *The physics behind Schrödinger's cat paradox*. National Geographic.
https://www.nationalgeographic.com/science/article/130812-physics-schrodinger-erwin-google-doodle-cat-paradox-science

Kumar, V. (2020). *14 best examples of convection with simple explanation*. RankRed. https://www.rankred.com/best-examples-of-convection/

Lanese, N. (2021, March 17). *Why does DNA spontaneously mutate? Quantum physics might explain*. Live Science.
https://www.livescience.com/quantum-physics-dna-mutations.html

Letzter, R. (2019a). *Giant molecules exist in two places at once in unprecedented quantum experiment*. Scientific American.
https://www.scientificamerican.com/article/giant-molecules-exist-in-two-places-at-once-in-unprecedented-quantum-experiment/

Letzter, R. (2019b). *The 12 most important and stunning quantum experiments of 2019*. Live Science.

https://www.livescience.com/most-important-surprising-quantum-physics-of-2019.html

Letzter, R. (2019c, May 20). *There's a brand-new kilogram, and it's based on quantum physics*. Live Science. https://www.livescience.com/65522-new-kilogram.html

Letzter, R. (2019d, June 28). *After decades of hunting, physicists claim they've made quantum material from depths of Jupiter*. Live Science. https://www.livescience.com/65827-metallic-hydrogen-claim.html

Lillelund, J. (2019, November 25). *Quantum computing explained so kids understand*. IBM Nordic Blog. https://www.ibm.com/blogs/nordic-msp/quantum-computing-kids-understand/

Lloyd, S. (2010). *Programming the universe: A quantum computer scientist takes on the cosmos*. Vintage Books.*thermodynamics?* Live Science; https://www.livescience.com/50941-second-law-thermodynamics.html

Lumen. (2019). *Planck's quantum theory: Introduction to chemistry*. Lumen Learning. https://courses.lumenlearning.com/introchem/chapter/plancks-quantum-theory/

Lumen. (2021a). *Applications of quantum mechanics*. Lumen Learning. https://courses.lumenlearning.com/boundless-physics/chapter/applications-of-quantum-mechanics/

Lumen. (2021b). *Discovery of the atom*. Lumen Learning. https://courses.lumenlearning.com/physics/chapter/30-1-discovery-of-the-atom/

MacDonald, F. (2019). *Scientists just unveiled the first-ever photo of quantum entanglement*. ScienceAlert. https://www.sciencealert.com/scientists-just-unveiled-the-first-ever-photo-of-quantum-entanglement

Macias, A. (2016). *The world's brightest scientific minds posed for this 1927 photo after historic debates about quantum mechanics*. Business Insider. https://www.businessinsider.com/solvay-conference-1927-2015-4

Mann, A. (2019, August 29). *What is the theory of everything?* Space; https://www.space.com/theory-of-everything-definition.html

Markoff, J. (2014, May 29). Scientists report finding reliable way to teleport data. *The New York Times*. https://www.nytimes.com/2014/05/30/science/scientists-report-finding-reliable-way-to-teleport-data.html

Masram, V. (2014). *Simple relativity: Understanding Einstein's special theory of relativity*. YouTube. https://www.youtube.com/watch?v=TgH9KXEQ0YU

Mathas, C. (2019, August 13). *EDN - The basics of quantum computing—A tutorial*. EDN. https://www.edn.com/the-basics-of-quantum-computing-a-tutorial/

Merali, Z. (2020). *This twist on Schrödinger's cat paradox has major implications for quantum theory*. Scientific American. https://www.scientificamerican.com/article/this-twist-on-schroedingers-cat-paradox-has-major-implications-for-quantum-theory/

Merriam-Webster. (2018). *Relative*. Merriam-Webster. https://www.merriam-webster.com/dictionary/relative

Metcalfe, T. (2021, August 18). *Fusion experiment breaks record, blasts out 10 quadrillion watts of power*. Live Science. https://www.livescience.com/fusion-experiment-record-breaking-energy.html

Minutephysics. (2011). T*he arrow of time feat. Sean Carroll.* YouTube. https://www.youtube.com/watch?v=GdTMuivYF30

Minutephysics. (2021). *General relativity explained in 7 levels of difficulty*. YouTube. https://www.youtube.com/watch?v=eNhJY-R3Gwg

MRI Lincoln Imaging Center. (2019). *Where was the MRI san invented.* MRI Lincoln Imaging Center. http://mrilincolnimaging.com/where-was-the-mri-scan-invented/

NASA. (2021). *Our Sun*. NASA Solar System Exploration. https://solarsystem.nasa.gov/solar-system/sun/in-depth/#:~:text=Orbit%20and%20Rotation

New Scientist. (2021). *Quantum physics*. New Scientist. https://www.newscientist.com/definition/quantum-physics/

Newton, S. I. (2019). *BrainyQuote*. BrainyQuote; https://www.brainyquote.com/quotes/isaac_newton_135885

Norton, J. D. (2020, November 9). *Origins of quantum theory*. University of Pittsburgh. https://sites.pitt.edu/~jdnorton/teaching/HPS_0410/chapters/quantum_theory_origins/

O'Connell, C. (2016, February 28). *Quantum physics for the terminally confused*. Cosmos Magazine. https://cosmosmagazine.com/physics/quantum-physics-terminally-confused/

Odenwald, S. (2019). *Gravity Probe B: Special and general relativity questions and answers*. Stanford. https://einstein.stanford.edu/content/relativity/a11758.html

Orzel, C. (2015a). *What has quantum mechanics ever done for us?* Forbes. https://www.forbes.com/sites/chadorzel/2015/08/13/what-has-quantum-mechanics-ever-done-for-us/?sh=43cb8c7b4046

Orzel, C. (2015b, July 8). *Six things everyone should know about quantum physics*. Forbes. https://www.forbes.com/sites/chadorzel/2015/07/08/six-things-everyone-should-know-about-quantum-physics/?sh=5eaab8af7d46

Pappas, S. (2019, August 28). *Quantum gravity could reverse cause and effect*. Live Science. https://www.livescience.com/quantum-gravity-could-scramble-cause-and-effect.html

Philo Notes. (2020, September 16). *Descartes's concept of the self.* PHILO-Notes. https://philonotes.com/index.php/2020/09/16/descartess-concept-of-the-self/#:~:text=Descartes

Popkin, G. (2018, April 25). *Einstein's "spooky action at a distance" spotted in objects almost big enough to see.* Science. https://www.sciencemag.org/news/2018/04/einstein-s-spooky-action-distance-spotted-objects-almost-big-enough-see

Real World Physics Problems. (2019). *Quantum physics for kids.* Real World Physics Problems. https://www.real-world-physics-problems.com/quantum-physics-for-kids.html

Rosengreen, C. (2019, April 16). *Researchers develop quantum device to generate possible futures.* HPCwire. https://www.hpcwire.com/2019/04/16/researchers-develop-quantum-device-to-generate-possible-futures/

Sarkar, D. (2021, May 19). *20 brilliant quotes from Albert Einstein, the theoretical physicist who became world famous.* Discover Magazine. https://www.discovermagazine.com/the-sciences/20-brilliant-quotes-from-albert-einstein-the-theoretical-physicist-who

Schrodinger, E. (1933). The fundamental idea of wave mechanics: Nobel Lecture, December 12, 1933. In *Nobel Prize.* https://www.nobelprize.org/uploads/2017/07/schrodinger-lecture.pdf

SchrödingerE. (1983). *My view of the world.* Ox Bow Press.

Skibba, R. (2018). *Einstein, Bohr and the war over quantum theory.* Nature. https://www.nature.com/articles/d41586-018-03793-2

Slocombe, L., Al-Khalili, J. S., & Sacchi, M. (2021). Quantum and classical effects in DNA point mutations: Watson–Crick tautomerism in AT and GC base pairs. *Physical Chemistry Chemical Physics, 23*(7), 4141–4150. https://doi.org/10.1039/D0CP05781A

Specktor, B. (2019, April 19). *This quantum computer can see the future — All 16 of them.* Live Science. https://www.livescience.com/65271-quantum-computer-sees-16-futures.html

Tate, K. (2013, April 8). *How quantum entanglement works (infographic).* Live Science; https://www.livescience.com/28550-how-quantum-entanglement-works-infographic.html

Technological Solutions Inc. (2021). *Physics for kids: Theory of Relativity - Light and time.* Ducksters. https://www.ducksters.com/science/physics/relativity_light_time.php

TEDx Talks. (2016, May 24). *Quantum physics for 7 year olds | Dominic Walliman.* YouTube. https://www.youtube.com/watch?v=ARWBdfWpDyc

The Editors of Encyclopedia Britannica. (2017). Thomson atomic model. In *Encyclopædia Britannica.* https://www.britannica.com/science/Thomson-atomic-model

The Royal Institution, & Carroll, S. (2020). *A brief history of quantum mechanics.* YouTube. https://www.youtube.com/watch?v=5hVmeOCJjOU

The Science Asylum. (2018). *The ultimate guide to space-time and relativity.* YouTube. https://www.youtube.com/watch?v=FdWMM6aXpYE&t=327s

Turner, B. (2021a, May 3). *Scientists make a breakthrough in developing the quantum internet.* Live Science. https://www.livescience.com/three-node-quantum-network.html

Turner, B. (2021b, August 16). *Famous Einstein equation used to create matter from light for first time.* Live Science. https://www.livescience.com/einstein-equation-matter-from-light

University of Chicago. (2019). *Machine learning reveals hidden turtle pattern in quantum fireworks.* Phys.org. https://phys.org/news/2019-02-machine-reveals-hidden-turtle-pattern.html

Valdano, D. (2017, March 3). *Physics mistranslated: Spooky action at a distance.* Medium. https://medium.com/physics-as-a-foreign-language/physics-mistranslated-fa83efdfa314

Watson, J. (2017, May 13). *Famous codes and ciphers through history and their role in modern encryption.* Comparitech. https://www.comparitech.com/blog/information-security/famous-codes-and-ciphers-through-history-and-their-role-in-modern-encryption/

Weisberger, M. (2019). *Truly spooky: How ghostly quantum particles fly through barriers almost instantly*. Live Science. https://www.livescience.com/65043-tunneling-quantum-particles.html

Westphal, J. (2019, August 8). *Descartes and the discovery of the mind-body problem*. The MIT Press Reader. https://thereader.mitpress.mit.edu/discovery-mind-body-problem/

Wigmore, I. (2019). *What is Schrodinger's cat?* WhatIs.com. https://whatis.techtarget.com/definition/Schrodingers-cat

Wilczek, F. (2016, April 28). *Entanglement made simple*. Quanta Magazine. https://www.quantamagazine.org/entanglement-made-simple-20160428/

Wolchover, N. (2014). *Quantum entanglement drives the arrow of time, scientists say*. Quanta Magazine. https://www.quantamagazine.org/quantum-entanglement-drives-the-arrow-of-time-scientists-say-20140416/

Wolchover, N. (2020). *Quantum tunnels show how particles can break the speed of light*. Quanta Magazine. https://www.quantamagazine.org/quantum-tunnel-shows-particles-can-break-the-speed-of-light-20201020/

Wood, C. (2021, June 10). *Physicists link "quantum memories" in early step toward quantum internet*. Live Science. https://www.livescience.com/quantum-internet-repeater.html

World Science Festival. (2015). *What is the difference between special relativity and general relativity?* YouTube. https://www.youtube.com/watch?v=_QA8y-xGlRI

Zych, M., Costa, F., Pikovski, I., Brukner, C. Bell's theorem for temporal order. *Nat Commun* 10, 3772 (2019). https://doi.org/10.1038/s41467-019-11579-x

Zyga, L. (2017, July 5). *Physicists provide support for retrocausal quantum theory, in which the future influences the past.* Phys.org; https://phys.org/news/2017-07-physicists-retrocausal-quantum-theory-future.html

Printed in Great Britain
by Amazon

55911832R00158